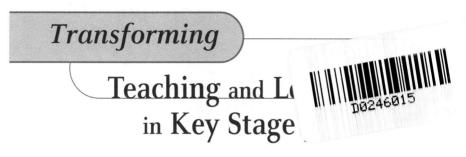

Transforming

Teaching and Learning in Key Stage

Science

Transforming

Teaching and Learning in Key Stage 3

Science

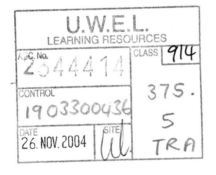
- Ken Mannion

- Marilyn Brodie

- Richard Needham

- Andy Bullough

Learning Matters

First published in 2003 by Learning Matters Ltd.

British Library Cataloguing in Publication Data
A CIP record for this book is available from the British Library.

ISBN 1 903300 43 6

Project management by Deer Park Productions
Cover and text design by Pentacor Book Design
Typeset by Pantek Arts Ltd, Maidstone, Kent
Printed and bound in Great Britain by Bell & Bain Ltd, Glasgow

Learning Matters Ltd
33 Southernhay East
Exeter EX1 1NX
Tel: 01392 215560
info@learningmatters.co.uk
www.learningmatters.co.uk

Contents

Raising standards, whilst keeping pupils enthused about science – is that not what all science teachers seek to achieve? Science education seems to be getting more and more attention, with recent reports on the subject from the Treasury (the 'Roberts' report) and the House of Commons Science and Technology Committee, as well as from the usual sources such as the Office for Standards in Education (OFSTED). The common theme in these reports is that youngsters, although achieving higher and higher scores in National Tests, GCSEs and A levels, are switching off from science. We believe that higher standards can also be achieved through motivating teaching and learning approaches. This book sets out some ways to bring together the 'standards agenda' with the developing 'motivation agenda'.

The authors have been involved for many years in developing motivating resources and activities for science teachers and their pupils, such as the Pupil Researcher Initiative (PRI), Researchers in Residence and the Inspire and Acclaim Projects. They have also provided teachers with a range of professional development programmes, most recently in information and communication technology (ICT), through the Science Consortium. We bring that experience to this book, as well as an understanding of the issues underpinning the science strand of the Key Stage 3 National Strategy. This book is designed to be used by teachers involved in implementing the Science Strategy. It is meant to accompany, and not replace the development work going on in schools to make the strategy work. Its principal aim is to support the continuing professional development (CPD) of teachers. We have tried to provide teachers with useful insights into certain aspects of many of the issues dealt with by the Strategy, but in a book of this length there are topics we have not been able to cover.

The Key Stage 3 National Strategy itself is providing schools with a range of materials and opportunities for professional development, but not all teachers will have first-hand access to all that is on offer. We have attempted to provide a resource that will be used alongside some of the more detailed advice which schools will be receiving. Although we have included a number of activities, we realise that teachers are likely to be working through the book by themselves. Most activities, therefore, can be done as individual tasks, although many will benefit from a team approach. As this is a book not a 'live' INSET session, we have provided extended descriptions or analyses on certain issues, which perhaps go beyond the input teachers might receive through taking part in a Key Stage 3 National Strategy workshop. In this way we hope to provide extra insight into some key issues.

The key purpose of teacher CPD, as promoted by the Key Stage 3 National Strategy and this book, is to transform and strengthen teaching and learning. This can best be achieved by promoting classroom approaches that motivate pupils and engage them more with their learning, as active participants. CPD should also promote more effective learning by addressing how pupils experience the curriculum – how progression is planned for and delivered, at the Key Stage 2/Key Stage 3 transition, and across the Key Stage and into Key Stage 4. For this to show results teachers need to have high expectations of their pupils and of themselves, so that the challenging targets set at national, Local Education Authority (LEA) and school level, as well as at the individual pupil level, can be secured. We see this book as providing a useful resource for teachers as they work towards these goals.

A model for teacher development

Readers of this book will be at different stages in their careers. Some of you will be students in training. Others will be experienced teachers, or teacher trainers and advisors. Wherever you are in your career, we hope that by reading this book, carrying out the activities, or using the book to underpin departmental INSET sessions, you will be able develop your professional practice.

Underlying the content of the book is a model for the development of teacher expertise, developed by the Centre for Science Education (Bevins, 2002) and based on the work of Brunner (in Berliner, 1994). It identifies four key stages in the development of an expert teacher: *novice, competent, proficient* and *expert teacher*.

Although we have not set the book out to explicitly address the needs of each group of teachers separately, we have tried to provide ideas and guidance which allow all teachers to benefit. We have done this by focusing on matters which allow the development of rational approaches, which can inform the novice's developing set of 'professional rules', provide competent and proficient teachers with alternatives to their current approaches, and give the expert teacher food for thought.

STAGE OF PROFESSIONAL DEVELOPMENT   NOVICE

Period in career
Student, and first one to two years of professional practice

Characteristics
The 'novice' teacher generally works from a set of context-free rules, is beginning to learn everyday classroom tasks and activities, and is beginning to identify elements common to the classroom and school environment.

Context-free rules: Berliner describes this issue by utilising an interesting analogy of learning to drive:

> When teaching a novice to drive we might tell them to shift from first gear at 12 mph despite the fact that such a rule is terribly inadequate for driving on hills, slippery roads or for certain engine ratios. An expert driver shifts when it is appropriate to do so or when the sound of the engine reveals to the experienced individual that it is time to shift.
>
> (Berliner, 1994)

This type of knowledge is rarely, if at all, displayed by novices and can only be acquired after years of experience. So, for this reason, we often provide novices with context-free rules to enable them to practice until they gain sufficient experience and know-how to adapt their practice. Novice teachers are given context-free rules such as, 'provide praise for right answers'. Novices tend to stick to the rules of teaching and remain somewhat inflexible as they experience and learn from those experiences the everyday mechanisms of teaching.

Learning: The novice classroom teacher has a limited range of tried and tested strategies and approaches and is, more than likely, unsure of their own model of teaching. Their training will have described various approaches, strategies and models, while their teaching practice placements allow for experimentation and trialing. Experience in the 'real' classroom environment provides opportunities to reflect on the approaches, strategies and models learned and to adapt and fine tune them. Responses from pupils, more established teachers and CPD activity add to the novice's basic, but growing, repertoire.

Identifying the environment: The classroom can be a frightening place even for the more experienced teacher. The novice teacher is beginning to identify the more obscure elements of the classroom and, indeed, the school. Novices will begin to gain awareness of such things as the dynamics of situations and pupils' expected classroom routines, as well as the classroom layout, ambience and equipment storage. They will also begin to gain an understanding of, for example, the school ethos, the school's standing in the community and management structures.

STAGE OF PROFESSIONAL DEVELOPMENT   COMPETENT

Period in career
Usually seen in teachers with between two and four years', experience. With further experience and a desire to develop and succeed most novices progress to competent practitioners.

Characteristics
Rational decisions: Competent teachers make conscious decisions and choices, through rational planning and prioritising. They will have specific, defined objectives and well thought out routes for achieving their objectives. From their previous teaching experience they will have developed a knowledge of what is important in the classroom and what is not – they will determine which events need attention and which can be ignored.

Timing: Because competent teachers have learned how to determine and prioritise classroom events they are less likely to make timing errors. They will have knowledge that enables them to make instruction decisions. For example, when to continue with a particular topic or activity and when to move on. They are beginning to recognise the differences in particular groups of pupils which helps their decision making based on specific groups.

Responsibility: A greater sense of classroom control leads to competent teachers feeling more responsibility for what happens in their classrooms – they are less detached than the novice. Competent teachers now begin to invest more of themselves emotionally in their teaching, recognising their successes and failures in a much more intense way than novices do.

STAGE OF PROFESSIONAL DEVELOPMENT   PROFICIENT

Period in career
Usually seen in teachers of five years' experience or more. Not all teachers will move from the competent to the proficient stage. The proficient teacher has acquired a broad base of experience, but more importantly has learned from that experience.

Characteristics
Intuition: Practice at this level has become faster, more fluid and flexible. Transition (changing tack) between topics is smooth and carried out through a sense of, or feel for, classroom events. Berliner describes this as developing an intuitive sense of the situation:

> Consider the micro-adjustments made in riding a bicycle or learning a dance step. At some point in learning to ride a bicycle or performing the mambo individuals no longer think about the kinds of adjustments needed. They no longer worry about balance and stop counting their steps to keep time to the music. In both cases they simply develop a more 'intuitive' sense of the situation.   (Berliner, 1994)

Pattern recognition: Through the accumulated broad base of experience at this stage, proficient teachers can now recognise similarities between classroom events and situations. For example, they will recognise intuitively that a certain approach taken failed for a similar reason that another approach taken previously did. Because they have experience of a wide range of cases proficient teachers can utilise this to inform their problem-solving strategies when they recognise an emerging classroom event which needs attention.

Analytic: While proficient teachers utilise intuition in the classroom they are still quite analytical in decision making. They will deliberate and carefully choose appropriate strategies.

STAGE OF PROFESSIONAL DEVELOPMENT   EXPERT
Period in career Expertise may be reached only after acquiring and learning from an appropriate amount of quality experience. However, as a guide many experts only reach the stage after at least eight years' practice.
Characteristics If the novice is deliberate, the competent performer rational, and the proficient performer intuitive, we might categorise the expert as often being arational. Experts have both an intuitive grasp of the situation and seem to sense in non-analytic and non-deliberative ways the appropriate response to be made. They show fluid performance, as we all do when we no longer have to choose our words when speaking. Experts appear to act effortlessly. Their performance is fluid and they move along without obvious careful calculation or deliberation until a problem occurs when they will employ a wide repertoire or database of experience. This repertoire enables them to quickly and smoothly analyse the problem and put into practice strategies to solve it. Expert teachers then, are much more rounded and holistic practitioners who do things that rarely go wrong, are unafraid of taking risks in their practice and are constantly looking to innovate.

The structure of this book

We have divided the book into two main sections:

Section 1 contains four chapters, each tackling a major issue for Key Stage 3 science teachers. The chapters are:

Chapter	Title	Key theme
1	Teaching, learning and assessment	Strategies for teaching and learning, target setting and assessment.
2	Scientific enquiry – what science is all about	Investigative skills, ideas and evidence and the nature of science. Interaction between investigative skills and ideas and evidence.
3	Science content: key ideas at Key Stage 3	Key ideas at Key Stage 3, and incorporating contemporary science and issues into schemes of work.
4	Raising standards	Strategies for more effective reviewing, planning and monitoring of pupil performance.

Section 2 is made up of six units, each based on one of the Qualifications and Curriculum Authority (QCA) schemes of work units for science at Key Stage 3. In these units we have revisited some of the ideas described in Chapters 1 to 4 in the context of six topics from the scheme of work. We have also added more in-depth descriptions of a number of issues, such as differentiation, literacy and contemporary science, again in the context of the six topics.

The six units are:

Title	QCA Scheme of Work reference
Cells	7A
Electrical circuits	7J
The solar system and beyond	7L
Atoms and elements	8E
Plants and photosynthesis	9C
Environmental chemistry	9G

To assist teachers unfamiliar with certain terms used throughout this book, we have included a set of simple definitions at the back of this book.

Whatever stage your career is at, we hope that you find the following chapters useful, stimulating and enjoyable to read and work through.

This section tackles four key areas for science teachers, with a chapter devoted to each. Chapters 1 (Teaching, learning and assessment) and 4 (Raising standards) relate to classroom processes, including assessment, data capture and processing. Chapters 2 (Scientific enquiry) and 3 (Science content) relate to curriculum content, as defined in the structure of the national curriculum for science.

We have taken some of the key themes that are being dealt with by the science strand of the Key Stage 3 National Strategy, and provided readers with a range of ideas and activities which will add to and in many cases go further than the support which the Strategy provides for teachers' continuing professional development.

The structure of each chapter is based on the following:

- general introduction to the chapter which describes the context in which the chapter is set, and a list of the topics covered;
- key issues in the chapter, based on the Key Stage 3 teaching and learning science standards, setting out the important learning points;
- interconnected sections which tackle important and relevant aspects of the overall chapter theme, including case studies and activities (for individual teachers or groups);
- summary, restating what you should have gained from reading the chapter and completing the activities.

Case studies have been used where relevant to exemplify some of the ideas described in the main text of the chapters. These are based on actual examples from schools. Activities have also been built into the chapters where we believe that teachers would benefit from stepping back from the text and looking at their own practice, or that of their department. The activities are aimed primarily at groups of teachers, particularly departments. By thinking through the issues we raise through the activities, we hope that readers will advance their thinking in a particular area, which should result in a positive effect on their practice. However, most of the activities can also be carried out by individuals, who may be reading and working through the text on their own.

Approaches to teaching, learning and assessment lie at the heart of this book. Looking critically at what teachers do with pupils during (and beyond) lessons, and how pupils respond, is critical to pursuing the two aims of raised standards and increased motivation.

We start by looking at some simple methods of target setting. Embarking on such a scheme within a department can at first seem daunting and time consuming. Having read this section we hope you will agree that the benefits of putting such a system into action, in terms of both standards and motivation outcomes, far outweigh the efforts involved in getting things started. We then go on to look at a range of ways to make science teaching more effective, in terms of the above two outcomes.

The topics covered include:

- firm foundations for learning: assessment and target setting;
- teaching and learning – making it work;
- from essentials to enrichment – interventions and motivational strategies.

KEY ISSUES

In this chapter, you will find out how to:

- set realistic but challenging learning objectives for teaching activities and use a range of methods, such as informal observation in class, short questions and assessment tasks, to assess pupils' learning in relation to these;
- use assessment evidence in short- and medium-term planning;
- use information from assessment to identify pupils' strengths and weaknesses and to provide feedback to pupils, helping them to reflect on their learning;
- use information from assessment to intervene effectively, adjusting teaching and learning activities;
- assess pupils in relation to expectations associated with a unit of work;
- make use of assessment findings to set individual pupil targets;
- have strategies for engaging all pupils in science and maintaining pupil motivation;
- identify appropriate teaching methods, including whole class, group and individual approaches, and use these effectively, making the best use of available teaching time and ensuring previous learning is recalled and new learning reviewed;
- use a wide range of activities well matched to the learning to be achieved.

Firm foundations for learning: assessment and target setting

Why is target setting such a powerful tool?

The power of target setting lies in the ability to review, evaluate and monitor development effectively as a means of improving pupil achievement. If used as a key element of your departmental Key Stage 3 Strategy, target setting can drive pupil improvement strategies as well as curriculum improvement strategies. At the level of the individual pupil, not only does target setting enable pupils to follow their own progress but it should also empower them to be realistic about their own true potential. This implies that pupils should track their own performance throughout the key stage. Data collected through target setting should also contribute to your departmental assessment information, and so classroom approaches should support the assessment model used in your department.

The table below shows some approaches that might form part of a common departmental system. This system should support and monitor the tasks and assessment methods of the course in use. It is based on the DfES Key Stage 3 tracking model that helps to meet the learning styles entitlement of pupils whilst also meeting the OFSTED specification for individual target setting and supporting teacher CPD (some schemes such as *Eureka* also offer strong

An academic pre-test	The purpose of this is so that pupils can form an evaluation of their own academic knowledge. It enables pupils to identify their strengths and weaknesses. The pre-test should be built upon work covered at Key Stage 2.
Evaluation of previous topic performance	This can be built into the same lesson as the pre-test. The aim of this is to review individual pupil progress not only through one topic but throughout a year and then an entire key stage.
The unit of work	This will provide the bulk of the course and is where the subject learning takes place.
Academic self assessment and progress tracking	Before the test this provides an opportunity for pupils to find their strengths and weaknesses within a given topic. This can be done through answering questions relating to specific areas within a unit of work. It should be possible to identify about four sub-divisions within any unit of work. These can be individually dealt with by progress tracking.
Self improvement tasks	These relate directly to the academic self assessment tasks. What we are suggesting is that pupils undertake revision tasks specific to their own needs.
Test	At the end of a unit of work.

Approaches to assessment and target setting

support packages that can form the framework for good target setting practice). The scheme of work should include the range of assessment and target setting approaches planned by the department.

The main thrust of time should be spent on the unit of work. The other aspects are designed to get the most out of that unit of work and should take no more than three lessons to implement.

Looking at the example, several key elements can be identified:

- regular testing;
- building upon positive experiences;
- opportunities for pupils to reflect upon their performance;
- suitable recording systems.

Collecting data to track individual pupil progress

The important thing is to be able to collect data that can be interpreted. Having a series of tests that are levelled against a series of marks makes comparing progress nice and easy. Each test is in turn marked out of the same total – in this way it is easy to compare the level achieved across a range of tests. Some published schemes provide ready-prepared tests. Self evaluation of performance, based on pupils considering their progress as defined by their test scores, is key to developing good reflective technique in pupils. Matched with target setting, self monitoring of performance then becomes an effective tool for raising standards.

Tracking progress with computers

If you have the data available, entering raw test scores for Key Stage 2 onto a spreadsheet could form the start of a pupil tracking system. Getting hold of the test grade boundaries from the report on that year's test would enable you to convert marks into targets, expressed as a sub-divided level, such as 5a. Progress

---

REVIEWING ASSESSMENT

- First choose a unit from your Key Stage 3 scheme of work.
- Make a list of all the assessment opportunities encountered by a group throughout the unit.
- Divide this list into formative (i.e. approaches which enable the pupil to improve their performance in the areas being assessed) and summative assessment.
- Repeat this for other units and review whether the pattern of assessment opportunities offered in these units follows a common pattern or style.
- Experiment with restructuring the pattern of assessment.

against these targets can then be monitored by entering 'levelled' test scores onto the spreadsheet. For more on this, see Chapter 4, Raising standards.

A clearly planned framework

In a busy science department the key element in making assessment work is planning: establishing the underlying framework, including resource and time requirements. Implicit within this chapter are the issues of curriculum progression and continuity, which are briefly described below.

- *Progression* is all about the individual moving onwards and forwards through building on existing skills and knowledge.
- *Continuity* is how we provide a learning experience that is not chaotic and fragmented but enables good progression to be maintained.

Progression and continuity are important not only during each key stage, but also at the boundaries between them. For teachers of Key Stage 3, a vital aspect of progression and continuity occurs at the boundary between Key Stages 2 and 3, where pupils in most areas actually change school.

Many secondary schools work with their partner primary schools on transition projects, such as the 'Moon Colony' (Cheshire L.E.A., 1996). This obviously requires thorough planning and co-operation between primary schools and the secondary school. Essentially the purpose of a transition project is to link the experiences of children from primary to secondary school. There are some clearly identifiable advantages of such a project to both the secondary and primary schools involved.

For primary pupils it can provide challenging, and meaningful work that can be both interesting and enjoyable, which in the process better equips pupils for their move into secondary school. The pupils bring with them an accessible record, which shows what they can do, enabling better progression to be made into Key Stage 3. Such transition projects also provide a better awareness for staff in the schools dealing with each key stage about what goes on in the other phase. Pupils begin work on the project in primary school, and then continue at the secondary school. Transferring examples of pupil work from transition projects adds richness to the data on Key Stage 2 assessment which will also pass to the secondary school upon transition. The two sources of data provide teachers of Key Stage 3 with a good starting point for target setting. The *Framework for teaching science Years 7, 8 and 9* (DfES, 2002) also provides a list of the scientific vocabulary that pupils should be familiar with by the end of Year 6, which should also help with the process of planning and target setting for Year 7 pupils.

Chapter 4, Raising standards, outlines some of the issues involved in curriculum planning, from short-term through to long-term planning. Here we will look at what planning needs to be carried out in order to make the scheme of work reflect the needs of your pupils, particularly at the vital time of transition from Key Stage 2 to 3 – a medium-term planning approach. For units that occur early in Year 7 we suggest developing an expanded scheme of work, giving an overview for the more detailed planning required for a unit, and for individual lesson plans. This need not be an arduous task but should be one that enables continuity to take place in pupils' learning experiences.

The elements that make up such an overview should include:

- pupil goals;
- teacher notes (including misconceptions);
- the learning intentions.

PUPIL GOALS

These are aimed at showing not only what we want pupils to achieve but also the teaching methods and resources available to enable pupils to have a planned learning experience. The goals from school to school can indeed be similar but there is space to put departmental flavour in here. Pupil goals may well be specific to your school. As such they can be an important starting point for planning a good scheme of work.

TEACHER NOTES (INCLUDING MISCONCEPTIONS)

Teachers should be reminded of the language that pupils should be familiar with when they come to secondary school. For example, in 'The Solar System and Beyond' (QCA scheme of work unit 7L) pupils should have already come across vocabulary related to this topic in their Key Stage 2 experiences during topics such as 'Light and Dark' (Y1), 'Light and Shadows' (Y3), 'Earth, Sun and Moon' (Y5) and 'Forces in Action' (Y6). The teacher notes section should also provide an opportunity to share expert advice and good practice across a busy department, for instance by dealing with common misconceptions, at both pupil and teacher level. Even at Key Stage 3 some subject specialists will feel uncomfortable about moving out of their own subject-based comfort zone. The aim should be to make positive, constructive and helpful comments that will help colleagues progress through what they may feel to be an unfamiliar topic.

THE LEARNING INTENTIONS

These should outline what learning you expect pupils to demonstrate by the end of the unit and should be based on a reworking of the 'Expectations' section of the QCA scheme of work, or local alternatives. It should set out what your expectations are under three headings:

- *all pupils will*: followed by learning outcomes beginning with active verbs such as *describe, show understanding, explain* and *recognise*. This indicates what even those who have made least progress should achieve.
- *most pupils will*:
- *some pupils will*:

The 'Moon Colony' transition activity mentioned above relates to Unit 7L in the QCA scheme of work science at Key Stage 3. We have used this to exemplify how we can document the points outlined above.

SCHEME OF WORK OVERVIEW   UNIT: THE SOLAR SYSTEM AND BEYOND (7L)

Pupil goals

The aim of this module is to open pupils' minds to big questions about our solar system whilst at the same time teaching pupils some key concepts about the Earth and its relative attributes in space and time in particular with respect to the sun and moon. Pupils should be able to describe our solar system and the Earth's position within it. They should understand night and day and why the seasons differ.

Pupils should have a chance to use a variety of media: Internet, CD-ROMs and specialist reference texts based in the Learning Centre. Through guided projects using these assorted media pupils will have the opportunity to communicate their findings via poster presentations. Their work will then be displayed in the class and Learning Centre.

This topic also offers considerable opportunity to develop concept models.

Teacher notes (including common misconceptions)

Pupils should have already met a range of vocabulary of use in this topic at Key Stage 2 including words relating to:

Light and dark: bright, dark, black, night, day, reflect, sun, darker/darkest, bright/brighter.

Light and shadow formation: transparent, opaque, shadow, block, direction, light travels.

Earth, sun and moon: sphere, revolve, orbit, spin, rotate, axis, sunrise, sunset, north, south, east and west, nouns and associated adjectives, (sphere/spherical), understand similar but distinct meanings e.g. rotate around, rotate on its axis, spin, orbit.

We should aim to build upon the above keywords and their associated concepts.

Care should be taken when describing the seasons. Winter is colder than summer because of the distance light rays have to travel through a tilted atmosphere that makes it colder at the Earth's surface.

The learning expectations

All pupils will:
- be able to explain the relative movements of the Earth, moon and sun
- be able to explain how the Earth differs from other planets
- know that the sun is a source of light as are the other stars
- know that the moon reflects light.

Most pupils will:
- relate eclipses, phases and seasonal changes to simple models;
- be able to explain the relative positions of the planets and their conditions;
- understand that planets and other objects in the solar system reflect light.

Some pupils will:
- be able to explain more complex models of the solar system using models, data such as distance from sun and orbital period;
- use appropriate order of number and compare the sun with other stars.

> **USING THE LANGUAGE THAT PUPILS HAVE ALREADY ENCOUNTERED IN A POSITIVE WAY**
>
> - Make a list of the key words already encountered in junior school by pupils at Key Stage 2 in a given topic.
> - In marker pen or large font make labels showing these key words.
> - Display the words in your classroom as a poster or hang then from your classroom ceiling.
>
> This should help pupils with progression and continuity and give them a readily visible vocabulary to refer to and work with. By putting the words above eye level pupils have to actively use them.

Teaching and learning – making it work

It's like choosing your favourite shirt!

We believe that the all-important starting point for good teaching is establishing good teacher–pupil relationships. The systems of target setting suggested above combined with individual teacher–pupil mediated goals provide a natural opportunity to establish a teacher–pupil dialogue. We also propose that keeping teaching and learning fresh and varied, although perhaps time consuming, will help your pupils learn and prevent you becoming jaded.

Imagine the classroom experience as being like looking in your favourite clothes shop for a new favourite shirt. When you enter the shop you may cast your eye over a wide variety of shirts of different size, colour, style and quality. Each of the shirts can represent one of a variety of teaching and learning approaches. Even though the variety is great each one relies on a hanger to hold it in place on the rail. The rail is the curriculum itself and the hanger represents the framework upon which your individual teaching styles can then be hung. So this part of the chapter splits into two sections: the essentials, and enrichment opportunities – or work shirts and 'going-out' shirts!

In distinguishing between essentials and enrichment, we are describing those teaching and learning approaches which you might use on a daily or weekly basis to take your pupils through the scheme of work (essentials), and those approaches you might use from time to time to enrich or enhance your pupils' experience.

The essentials

Before embarking upon this section it is important to realise that the Key Stage 3 Strategy for science assumes a certain degree of capability in the teachers who are taking part. By capability we mean confidence in the processes that enable a teacher to engage pupils in a rewarding way. It is worth checking what some of these key processes are. In terms of our categorisation of teacher expertise, the Key Stage 3 Strategy seems to be assuming that teachers are working in the 'competent' to 'proficient' areas, although the Strategy does not make this distinction explicitly, and obviously has something to offer teachers from 'novice' through to 'expert'.

We believe that the keys to competency and proficiency are consistency and familiarity with your own teaching environment. Consistency is the basis of forming strong positive relationships with your pupils. It includes consistent marking that should include defined areas such as content presentation and effort, which should set pupils targets for how to make more progress even if only by saying 'keep up the good work'. Good teachers develop consistent rewards and sanctions that in themselves enable the teacher to better motivate a class or an individual within that class. Rewards can be simple. Verbal praise is key to helping pupils make good progress. Establishing good communication with parents can also be key to making progress. Phone home good news as well as the 'less than good'. This helps to establish a teacher–pupil–parent relationship that can become very powerful when you want to engage parental support.

Positive contact with parents can be written as well as verbal. Encouraging pupils to collect merit certificates or stamps is an easy way to create a simple yet powerful positive reward system. A sample combined certificate and letter home is shown below. This could be used when a pupil receives ten merits in a given topic.

The key elements to this letter are those highlighted in bold – these are the aspects which make the letter personal. It is relatively easy to create such a letter as a template on a computer and easy to change personal details. Please note that no reference to the pupil's gender has been made – this makes the task even easier when changing names. Why not incorporate a nice bit of clip art into your A4 template that has some relation to either success or science? If you feel ambitious why not create a template for each of the units that you teach.

Another key to competency and proficiency is good planning. As you move through your career the scope of your planning will change. As a novice or teacher in the early stages of your career, planning is likely to revolve around lesson planning (including ordering laboratory equipment from the technician). In many schools new teachers are also involved in departmental planning processes such as contributing to schemes of work, although responsibility for this is likely to become greater as you move through your career.

Being totally familiar with and being in control of your teaching environment is also an aspect of being proficient. Although many school laboratories have fairly inflexible furniture layouts, many provide opportunities for customisation. Depending on your teaching aims, there may be different arrangements of the desks or benches which allow you to adopt different

6/3/02

Dear **Mrs Stevens**,

I am writing to you to tell you about the good effort that **Andrew** has put into science lessons.

During Year 8 **Andrew** has worked enthusiastically in class to gain 10 merit stamps for well-completed pieces of work. Through a variety of activities **Andrew** has looked at the following key areas of science:

- the biological sciences;
- the chemical sciences;
- the physical sciences.

Andrew has found out about variables and how to control them. These, along with data handling skills, form a basis for scientific thinking. If **Andrew** continues to proceed well in science you will undoubtedly be hearing from me in the future. Well done **Andrew.**

Yours sincerely,

A C Bullough

(Science Teacher)

teaching approaches. How well suited is your classroom layout for each of the following:

- small group discussion;
- multimedia and resource-based work;
- whole class discussion.

Good timekeeping is also key to science teaching – and pacing a lesson well leads us nicely into the Key Stage 3 Strategy guidance on teaching itself.

REVIEWING YOUR CLASSROOM LAYOUT

- Draw a plan of your classroom as it currently is set out, on squared paper.
- Experiment with different plans, setting out the furniture in different layouts which enable, for instance, better group work or access to multimedia resources.
- Choose a layout that you think will work (tip: try to move yourself out from behind that big desk) and try it in your classroom.

Interactive teaching

The *Framework for teaching science* sets out an approach which encourages teachers to adopt, 'direct interactive teaching' as the key methodology. This is described as being based on engaging pupils actively, as a whole class, through discussion and questioning, collaborative working, and explaining their work to others. Lessons should be tightly structured and have clear objectives. Each lesson should be divided into three sections:

- starter activity;
- main activity;
- plenary activity.

The bulk of the average lesson would be spent on the main activity, with starter activities taking perhaps 5–10 minutes from a 50–60 minute lesson, and the plenary taking up a similar amount of time. The *Framework* describes how this approach can be used flexibly, sometimes having several plenaries in the course of a single lesson. It also describes a range of approaches to teaching, under the following headings.

APPROACHES TO DIRECT, INTERACTIVE TEACHING	DESCRIPTION
Directing and telling	<ul><li>Objective sharing</li><li>Highlighting important points</li><li>Instruction giving</li></ul>
Demonstrating	<ul><li>Practical work</li><li>Visual displays including models, projected microscopic views, data projectors and electronic whiteboards</li></ul>
Explaining and illustrating	<ul><li>Explaining links between evidence and conclusions</li><li>Using models and analogies</li></ul>
Questioning and discussing	<ul><li>Maximising pupil participation and inclusion</li><li>Open and closed questions</li><li>Listening and responding to pupil answers</li><li>Encouraging thinking</li></ul>
Exploring and investigating	<ul><li>Supporting the development of investigative skills</li></ul>
Consolidating and embedding	<ul><li>Practising and developing new skills and understanding</li><li>Reflecting on learning</li><li>Encouraging transfer of learning across the curriculum</li></ul>
Reflecting and evaluating	<ul><li>Picking up on pupils' errors and misconceptions</li><li>Evaluating pupils' work and presentations</li></ul>
Summarising and reminding	<ul><li>Reviewing learning</li><li>Restating expectations and reminding of key points</li><li>Linking to the next stage of learning</li></ul>

We believe that a key element of all this is engaging pupils actively in the learning process. Whilst the above advice sets out a framework few would argue with, we also need to consider the mechanics of achieving this necessary active involvement.

Many of you will be familiar with a wide range of teaching and learning approaches, including role play, small group discussion and pupil presentations, as well as more didactic approaches such as whole class exposition and demonstration. However, we feel it would be useful to describe more fully some of the approaches that could be used to help create the interactive learning environment described in the *Framework*. (Some examples of using contemporary science and issues to produce starter activities are given in Chapter 3).

Smarter starts and powerful plenaries

Lessons should start in an engaging way that will get pupils quickly on task. There are many ways to do this. Here are a few ideas:

GO FOR 5

This approach assumes that the aim for the lesson is clearly displayed to the class. With *Go for 5* pupils have to write down five pieces of information they know about one (or more) of the key words within the lesson's aims. You identify the word that they are to focus on. The pupils can then share their five examples with a partner or embark on whole class discussion.

CONNECT

Pupils start the lesson by writing a brief sentence to describe what they learned from the previous session, then share or discuss.

SUGGESTIONS

Pupils are asked to write down what they would like to understand better from the lesson they are just starting. This works best if it is not the first lesson on a topic.

HANDS UP

Ask pupils a short series of questions. They have to register their answers by putting up their hands (or not). For example, 'How many people think that there is life on Mars?' Questions can range from factual ones through to those requiring an opinion. Record pupil answers, and return to them later during the plenary session to see if any pupils have changed their minds.

These are brief activities aimed at engaging the pupils quickly with the lesson content and getting them quickly on task.

These simple starters to lessons can also be readily applied to the all-important plenary part of your lesson. Leave enough time to round off the lesson properly, and don't just tell pupils what they have learned. Ask them. For example, use a 'Go for 5' activity. Get pupils to list five things (or three or seven) that they have found out or understand better. Remind pupils of what the objectives of the lesson were, and, whatever activity you decide upon, one aim must be to review whether the objectives have been met. Powerful plenaries consolidate what has been learnt, point the way to the next lesson and involve pupils in actively contributing.

Main lesson activities – active teaching and learning approaches

Many lessons will have practical work as the main activity. Others will have teacher explanation, whole class questioning, demonstration and discussion as key elements. The *Framework* describes a range of types of direct, interactive teaching, which is summarised here on page 18. Below we describe in more detail some of the approaches which look more at what the pupil is doing – shifting the focus slightly from how the *Framework* describes things. These approaches are taken from the range described in *Active Teaching and Learning Approaches in Science* (Centre for Science Education, 1991).

SMALL GROUP DISCUSSION USING AGENDAS

The essence of this approach is to make the discussion purposeful so that it helps your pupils explore ideas and encourages active participation in their own learning. It also allows you to keep close control and is particularly useful for pupils with less experience of more open, free ranging discussion. The first step is to prepare a structured agenda for pupils to follow (see below). Start with concrete questions or items and lead on to more abstract or evaluative items. The finished agenda should move pupils in steps from more factual discussion, perhaps involving recall, to the application of knowledge, evaluation and speculation. The discussion can be widened to the whole class by getting pupils to present ideas to each other, or by combining groups in a 'snowball'. By giving each group an agenda, you will be able to move between groups, helping to move the discussion on. You will also be able to judge the level of pupil understanding as you listen to the discussions. You will also be able to encourage pupils who are less forthcoming to contribute to the discussion in a less threatening forum than whole class discussions. An agenda-based discussion can be used as a link between the lesson's main activity, such as an experiment, and the plenary session at the end. Having spoken in their small group, pupils will gain confidence and be more willing to contribute in the whole class setting.

JIGSAW ACTIVITIES

This method, also based on agendas, provides pupils with opportunities to collaborate in a structured way. It can be used as a means of tackling a short topic where there are several sets of information to deal with. It is based on the 'home group – expert group approach' and works best in classes where pupils are used to working in fairly stable groups of four.

Small group discussion agenda – acids and alkalis

IN PAIRS

- Tell each other what results you recorded in the indicator experiment.
- Talk about the colour changes you observed when you added the chemicals to the indicators.

(3 minutes)

IN FOURS

- Share your information with the other pair.
- Is there a pattern in the results?
- Can the indicators be used to group the chemicals into different categories?
- Use the terms *acid*, *alkali* and *neutral* to describe the groups.

(5 minutes)

AS A WHOLE CLASS

Decide why indicators might be useful substances.

(5 minutes)

Each home group of four is set the same task, via an agenda or whole class discussion. The task is then divided into four components. Each member of the home group then has responsibility for dealing with one of the four aspects. A typical example could be looking at the origins and uses of a range of renewable and non-renewable energy sources. The home group has to produce a presentation for the class, using posters, overhead projector or presentation software such as *Powerpoint*. Each member of the home group then joins up with three member of other groups who have the same energy source to research. This is the 'expert group'. Pupils then carry out the research activity, using the available resources, and produce a summary of their findings. The expert groups then disperse, and the reassembled home groups use the information from each of the four 'returning' experts to produce their presentation. Your role will be to circulate amongst the groups moving things forward, clarifying the task or the science where pupils need encouragement or explanation.

A jigsaw activity can take up the main part of a lesson, and so it is essential that the starter activity provides pupils with enough understanding of the task to carry it out effectively. It would be advisable to allocate longer to the plenary session following on from a jigsaw activity, as you will need to build in time for groups to feed back, and for you to summarise and remind pupils what they have learned.

ROLE PLAY AND SIMULATION

Our experience of working with teachers and suggesting that they try role-play activities is that many people are reluctant to try what they see as a risky approach. We are often told, 'I tried it once and it didn't work', although many of you will be good at running such activities. What follows is a short description of how to structure a role play, following through the various stages, from preparation through to debriefing and follow-up work. Most of the problems teachers face in running role plays centre on the fact that they are not set up effectively – but just launched on an unsuspecting class.

What do we mean by 'role play' and 'simulation'? Many of you will be familiar with computer simulations. That's not what we are describing here. Role plays are person centred, in that issues and concepts are explored by asking pupils to imagine that they are in someone else's shoes. We are using 'simulation' here to describe a similar type of approach, but where the activity is job centred, where the emphasis is on carrying out certain tasks and pupils behave according to the task in hand.

Simulations provide a framework in which pupils are active participants and are given responsibility for their actions and decisions within the overall simulation scenario. Role plays provide a similar framework, but are used where you want pupils to explore different points of view, to challenge stereotypes and attitudes.

Both types of activity are, when run well, highly motivating for pupils and provide rapid feedback to both pupil and teacher on what is being learned.

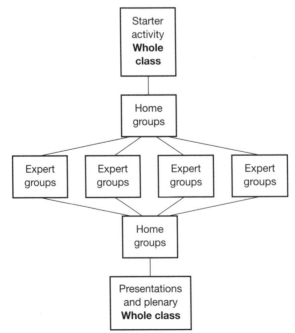

The jigsaw approach

As with a jigsaw activity, a role play or simulation can take up the main part of a lesson. The starter activity can be used to set the scene and to begin preparation for taking on the role. The plenary session can be used for debriefing, consolidation and review. The five stages of a role play or simulation are:

1. Preparation

- Be fully aware of the content and requirements of the activity, particularly if it is from a published resource.
- Select any stimulus material you might use.
- Make sure that the pupils have the required knowledge from earlier lessons to participate fully.

2. Briefing – getting into role

- Introduce the activity and stimulate pupil interest so that they take part fully.
- Explain the structure of what you are about to do.
- Identify who will take on what role, and group the class accordingly.
- Introduce pupils to the role cards or other items which provide them with insight into the role they will be playing. This can be done with the whole class, with the role groups, or by individuals studying the material relevant to their own role.
- Pupils with the same role could support each other by working together to analyse their role, making note of key points they have to get across or tasks they have to perform. This way, they can help to 'work up' the characters.

3. Running the action

- If the briefing was effective, pupils should be fully conversant with the situation being dealt with, including the science content, the basic feelings, beliefs and attitudes being represented, or the scope of the task being simulated, and their place within it.
- Once it is underway, step back a little and observe what is going on when the pupils are engaged in the discussion or task.
- Co-ordinate any changes in the activity as it unfolds, such as when pupils have to take turns to make inputs to the role-play discussion.
- Ask any questions in role – have you got a defined role for yourself?

4. Debriefing

- This is usually best begun with pupils still in role.
- Upon completion of the main activity, pupils should discuss in their role-play groups what they have learned through the activity, in terms of new knowledge and understanding, changed attitudes or greater insight. This could lead to a whole class discussion.

REVIEWING TEACHING AND LEARNING APPROACHES

- Fill in this grid (or a copy) to identify which teaching and learning approaches are used in your department. Alternatively, it can be completed by individuals and then compiled into a summative grid.

Teaching and learning approach	Which do I (or the dept.) use now?	How frequently is it used?	Do we need to increase or decrease the use?
Small group discussion			
Jigsaw			
Role play			
Whole class question /answer			
Drama			
Simulation			
Visits			
Visitors			
Science events outside school			
Demonstration			
Class note-making from texts/resources			
Science games			
Use of ICT for monitoring			
Internet			

- Add other approaches to the chart.
- Does the chart show that there is a consensus for the department to develop a wider range of teaching and learning approaches?
- What are the implications for staff CPD?

5. Follow up

- Pupils can now come out of role.
- Remind pupils of what they have learned in the activity. When dealing with controversial issues (see Chapter 3) it may be relevant to agree with pupils what learning has actually taken place.
- Emphasise and reinforce any science learning that has underpinned the activity.

From essentials to enrichment – interventions and motivational strategies

The task is to produce a changed environment for learning – an environment in which there is a new relationship between pupils and their subject matter, in which knowledge and skill become objects of interrogation, enquiry and extrapolation. As individuals acquire knowledge, they should also be empowered to think and reason.

(Glaser, 1984)

Intervention is a process in which the teacher uses some specific methodology to improve the pupils' learning experience. Intervention strategies aid pupils' learning and they are often linked to the notion of thinking skills and personal capabilities (which includes key skills). Most intervention strategies are based on the idea that by helping a pupil to develop a specific skill better, we will achieve improved pupil motivation and success. All this should be true of all classroom activities perhaps, but here we will consider two formalised intervention approaches: 'personal capabilities' and 'CASE'.

Personal capabilities

Personal capabilities (PCs) are a set of generic capabilities considered to be influential in the social, academic and professional lives of individuals. They are based on the notion of multiple intelligences (Gardner, 1993) and link to other aspects of learning such as accelerated learning, thinking skills and emotional intelligence. The list of PC-related skills presented here is non-exhaustive and under continual review. However, it provides a basis from which to begin considering their importance in assisting broad-based learning.

PERSONAL CAPABILITY	DESCRIPTION – BEING ABLE TO....
Positive self image	value oneself and one's achievements
Self management	take charge of one's own learning
Creativity	think and share new or novel ideas
Verbal communication	communicate one's knowledge, understanding, opinions and feelings appropriately
Critical thinking	critically review and evaluate practice in order to improve
Self motivation	motivate oneself to do what needs to be done
Problem solving	analyse a problem and form strategies to work towards a solution
Tenacity	stick to a task in order to meet deadlines
Teamwork	work well in teams
Social intelligence	respond appropriately to different situations and people

The development of personal capabilities can be based on awareness raising exercises where pupils are encouraged to reflect on and develop their skills such as team working, tenacity or communication. Science is a great vehicle for this approach because it promotes skills which are at the heart of what doing science and scientists is all about – good scientists have well developed PCs.

Research (Bianchi, 2002) into the development of personal capabilities in the Key Stage 3 science curriculum has explored the methods and implications of explicitly addressing the development of key skills and characteristics such those listed in the above table through the science curriculum. We have found that the benefits to pupils are:

● improved self awareness;

● improved confidence and behaviour;

● improved lesson structures;

● emphasis on pupil-centred teaching and learning methods and project work;

● relevant and person-centred self-evaluation processes;

● enhanced opportunities for the development of teamwork, creativity and problem solving abilities.

The PC approach can expand the scope of planned learning objectives. Rather than just focusing on subject knowledge and procedural understanding, by addressing objectives related to the development of PCs we can develop a wider range of learning outcomes. Whilst this is an important 'added' dimension to science, it also provides a vehicle and rationale for the development of cross-curricular learning. For instance, the introduction of citizenship as a statutory subject is involving science teachers in dealing with new areas of content, or old areas with a different approach. If PCs can add an extra dimension, the methodology for developing these skills is rooted in the active learning approaches described earlier, and the individual target setting processes described at the beginning of the chapter.

USING THE PERSONAL CAPABILITIES APPROACH

This will need some preparation but should provide a method by which pupils can make smarter starts to their lessons whilst developing their own personal capability skills.

● First identify a number of key skills which pupils could use in their lessons. They might include the following.

COMMUNICATION

Target	Date
To share my opinions with others	
To change the way I speak depending on the people I am with	
To accept other people's views	
To justify why I hold my opinions	
To discuss topics with other people	
To plan what I say so that my ideas can be understood by others	
To listen and respond well to other people	
To seek advice when necessary	

TEAMWORK

Target	Date
To co-operate with other members of the team	
To help decide what needs to be done in the team	
To help decide who will do what in the team	
To discuss issues with other team members	
To take responsibility for work in the team	
To offer suggestions for teamwork	
To help reach an agreement with others in the team	
To be willing to critically evaluate ideas	
To be willing to change if necessary	
To check what other people are doing in the team	

- Display the PCs in your classroom.
- At the start of a topic or lesson ask pupils to identify which two of these capabilities they would like to develop more, individually or in groups.
- Pupils write their choice at the start of the lesson then review their progress on this at a later date or at the end of the lesson or topic.

The development of PCs can also be seen to add value when used in conjunction with a structured 'thinking tool' approach. CASE will be discussed a little later. Another example of a systematic method to support broad-based learning is GRASP®, or 'Getting Results and Solving Problems', developed by the Comino Foundation. The GRASP® approach is a straightforward system that can be applied to almost all aspects of the teaching and learning experience. The approach is based on getting pupils to work through a set of questions prior to beginning a task. The approach is used in the guidance material for CREST Awards projects and in many aspects of the Pupil Researcher Initiative.

THE GRASP APPROACH	
Stages	*Questions*
Purpose	• What is it that I (or we) really want to achieve?
Criteria	• How will we know when we have achieved it?
Alternatives	• What possible ways can we use to get the results?
Selection	• Which one best satisfies the criteria?
Monitor and control	• How will we know whether we are keeping the process on track and what action to take if we are not?
Review	• Then we act, reviewing continually our purpose, criteria and progress, finally reviewing the result.

CASE

In science one of the most popular forms of intervention strategy is CASE, or Cognitive Acceleration through Science Education (Adey, Shayer and Yates, 1995). CASE is a scheme developed by Kings College, London with the aim of accelerating the development of a series of thinking skills through structured experiences, or interventions, in science lessons at Key Stage 3. Schemes such as CASE have been shown to work in improving the performance of pupils at GCSE not only in science but also in other core subjects. However, CASE can be hard to implement by a teacher working in isolation within a department.

Small-scale piloting of ideas in schools can lead to longer-term curriculum changes. However, in order to establish CASE fully, it needs to be embraced across a whole department. This may be problematic in some cases due to the prescriptive nature of such schemes and the occasional inertia of some colleagues who resist change (perish the thought). However, if you are on your own, give it a try. You may convince others to join you as you begin to demonstrate the benefits. If you do intend to give CASE a try we recommend using the test on 'Volume and Heaviness' followed by activities 1–5. This provides as good a starting point as any for any eager test pilots reading this. You will find that doing CASE activities will also support the development of scientific literacy and underpin the development of investigative skills.

Strategies for motivation

Strategies for raising pupil motivation are not merely for use with pupils who may have lost interest in science. They can recharge even the most inspirational teacher's batteries, and motivate even the keenest pupil enthusiasts to greater things. However, although they are suitable for all pupils, particular groups may benefit in dramatic ways. Groups of pupils worth targeting with enrichment and enhancement opportunities include:

- disaffected boys and girls (who may be awkward to teach);
- gifted and talented underachievers (who may be very frustrating to teach);
- low ability groups (who may often be a challenge to teach).

Enrichment and enhancement opportunities are not only worthy activities but they also gain recognition for those who use them. Departmental colleagues could be surprised at how well motivated your pupils become following one of these activities. OFSTED also recognises their importance, and many of their reports cite examples in positive terms. So what are we talking about? Here are some examples.

SCIENCE CLUBS

- The BA (British Association for the Advancement of Science) offers support and resources to schools and science clubs through the BAYS (BA Young Scientists) awards scheme. This includes the First Investigators awards scheme for five- to eight-year-olds and the Young Investigators for 8- to 13-year-olds. Many schools use BAYS as the basis of science clubs. The BA also runs the CREST Award scheme, which provides an accreditation system for pupils carrying out project work in science and design and technology. Many CREST projects are also carried out in science clubs, although some are done in lesson time.
- The BA also organises BAYSDAYS, which are children's science festivals held at different venues around the UK each year. Many of the participants belong to science clubs back in school. Events feature:
  - workshops, dramas and lectures with lots of ideas to take away;
  - hands-on activities where children 'do' science guided by experts;
  - a unique, exciting atmosphere created by the large number of enthusiastic visitors.
- The Science Discovery Clubs Network was launched by the BA as part of Science Year. It aims to build a national network of science clubs of all shapes and sizes which can interact, swap ideas and participate in regional events and national mass-participation activities. The network is open to all science-based clubs throughout the UK.
- Young Engineers Clubs are the national network of science, engineering and technology clubs based in schools and colleges throughout Britain.
- Astronomy clubs exist in many schools, usually where there is an enthusiast teacher.

SCIENCE FAIRS AND CONFERENCES

- Express Yourself pupil science conferences, part of the Pupil Researcher Initiative, are inspiring events for youngsters. Although PRI is aimed primarily at Key Stage 4, many Key Stage 3 groups attend. Five regional events are held around the UK each year, with a UK-wide finale at the famous Royal Institution. Groups of pupils make seminar presentations to other pupils and invited adults such as scientists and engineers. Other pupils mount displays for the poster session. These events are excellent for motivating youngsters and for showing them the importance of communication in science. But then we would say this, since we organise them!

- Chemistry at Work events, organised by the Royal Society for Chemistry, present to young people some of the principles of chemistry as they are applied in industry, research and everyday life. Event are held at many venues around the country each year.

- Physics at Work events, organised by the Institute of Physics, are similar to the Chemistry events, but obviously focused on physics.

- The UK-wide network of Setpoints organises many events locally.

BRINGING SCIENTISTS AND ENGINEERS INTO THE CLASSROOM

A number of schemes exist which enable schools to bring scientists and engineers from industry or academic research into the classroom.

The Science and Engineering Ambassadors Scheme (SEAS), funded by the DTI and organised locally by Setpoints is designed to provide schools with access to 'Ambassadors'. SEAS is now the home for the popular Neighbourhood Engineers scheme, which many schools already participate in. A number of other existing national schemes are affiliated to SEAS. These include:

- Researchers in Residence – which places young, positive role models into schools, from the ranks of the country's PhD science and engineering students. The scheme is funded by the research councils and the Wellcome Trust, and is co-ordinated by us at the Centre for Science Education. Well worth a try!

- BAE Systems' Ambassadors. This is similar to researchers in residence, but with BAE Systems employees. Most placements take place within commuting distance of BAE sites.

Check with your local Setpoint what other organisations provide scientists and engineers in your area.

SCIENCE CENTRES

There are now over 50 science centres (including some industrial museums which have science galleries) across the UK. UK Magna (Rotherham), the Eden Centre (Cornwall), the National Space Centre (Leicester), and @Bristol (Bristol) are amongst the more recent additions.

MOTIVATION AND ICT

Many pupils respond positively to the opportunity to use ICT in science lessons, although the same is not true of all science teachers! The problems of organising access to computers and uncertainty about lesson outcomes mean that this powerful strategy still carries too great an element of risk for some. For others, access to new gadgetry such as laptop computers, interactive whiteboards, digital cameras and microscopes has generated much enthusiasm for trying out new approaches.

Motivating factors identified by teachers include the following:

- Using a digital projector to show simulations of difficult concepts or tricky practical work is an effective strategy that enlivens whole class teaching.
- Organising pupils to work at computers singly or in pairs gives you more opportunities to work with small groups, supporting those that need help or extending faster workers.

Motivating factors identified by pupils include the following:

- Allows experimentation in a safe environment.
- Gives pupils more control over the pace of the lesson.
- Enables work to be produced to a high standard of presentation.
- Presents information in visual and auditory forms as well as text.
- Provides rapid feedback to pupil responses.
- Frequently presents an element of challenge.

INSPIRING RESOURCE MATERIALS

As well as text books and Key Stage 3 published schemes, a wide range of curriculum materials, specially designed for the Key Stage 3 curriculum, have been produced by industry, research councils, the learned societies and other public bodies. These often provide motivating contexts which put science into a real world situation. Good examples are:

- *Engineering our Future* (BAE Systems) – flight-based resources (hard copy and web-based) which fit the Key Stage 3 scheme of work;
- *Acclaim* (the Royal Society) – video and teaching pack looking at the life and work of leading living scientists;
- *Corus* – a wide range of materials science (particularly steel) resources;
- *BNFL* – a range of resources focusing on energy.

Other major producers of Key Stage 3 resources include:

- NERC (Natural Environment Research Council)
- BBSRC (Biotechnology and Biological Sciences Research Council)
- MRC (Medical Research Council)
- BP

- Shell Education Service
- BT
- National Dairy Council
- The Association of the British Pharmaceutical Society
- Glaxo Wellcome
- Esso
- Wordwide Fund for Nature.

Summary

In this chapter we have looked at a wide range of issues related to teaching, learning and assessment. We began by looking at how we can support pupil learning through a planned and well thought out approach to pupil monitoring and formative assessment. We also considered the importance of planning for continuity across the Key Stage 2 / Key Stage 3 transition.

In summary, this chapter has highlighted the need to:

- produce a target setting and assessment plan for the units in your scheme of work;
- plan for effective continuity and progression across the Key Stage 2 / Key Stage 3 transition;
- select and use appropriate teaching and learning approaches based on the objectives of each lesson;
- use a selection of enrichment activities to extend your pupils' science experience.

Teaching expertise is developmental. To fully engage with the requirements of the Key Stage 3 Strategy teachers should be aware of the base-line level of competency or proficiency which the Strategy presumes, although teachers at all levels of expertise can benefit in some way from the process. Teaching could be seen to be a balance between making sure the 'essentials' are carried out effectively, whilst building in opportunities for enrichment wherever possible or desirable. Direct and interactive teaching should include approaches that focus on what the pupil is doing as well as what the teacher is doing. Active teaching and learning approaches can motivate and help to secure effective learning. Curriculum enrichment activities can help to transform your pupils' experience of science – helping them to form a positive view of what science is all about.

Having spent time on this chapter we hope that you will have been able to nod your head with some parts of it and agree 'yes I do that' or 'I agree with this idea in principle'. However, we feel that we would not have achieved our goal if there were nothing new for you here or nothing worth giving a try. Science teaching is by its very nature an experimental process so we hope that you will experiment with some of the ideas here by trying to incorporate them within your own teaching experiences.

Further reading

Adey P S, Shayer M, and Yates C, 2nd Ed (1995) *Thinking Science: the curriculum materials of the CASE project.* London. Thomas Nelson and Son.

This pack contains all the curriculum materials necessary to run CASE in your school. The scheme is best run if you also go on CASE training, which can be organised by contacting the School of Education, Kings College, London.

Centre for Science Education, (1992) *Active Teaching and Learning Approaches in Science.* London. Collins Educational.

A handbook containing chapters on a wide range of teaching and learning approaches, including small group discussion, role play, problem solving, and active assessment.

This chapter looks at how we teach pupils about doing science, and how we teach them *about* science. However, most of the time we spend in the school laboratory we are teaching pupils science facts and concepts, and we need to consider how we shift the balance towards the two aims listed above. This is the purpose of a recent change in the science curriculum, that is, the introduction of Sc1 (scientific enquiry).

Scientific enquiry lies at the heart of science, both in schools and in the world of the research laboratory. Scientists are involved in enquiries that extend the reach of human knowledge about the biological and physical world. School pupils are taught science, a selection from this developing area of human knowledge, and later tested on much of it. But they also need to be taught about how science works, its power, its limitations, its social context and its methods; about how we extend this human knowledge. Without this dimension, science content would be divorced from its source and significance. It is this link between the content of science, and how we know the content in the first place and its consequences and implications, which makes scientific enquiry a vital part of the National Curriculum.

The topics to be covered include:

- the structure of Sc1 (scientific enquiry) – two halves of the same coin?
- ideas and evidence in science;
- investigative skills.

KEY ISSUES

In this chapter, you will:

- learn about some of the common misunderstandings pupils may have about the nature of science and experimental evidence;

- consider how science is based on an interaction between evidence and representations or models of the world which offer explanations;

- meet a range of ways of investigating scientific questions so that you can choose ways of developing pupils' competence in investigations;

- recognise that pupils' competence in investigation, including formulating and testing ideas, needs to be developed explicitly.

The structure of Sc1 (scientific enquiry) – two halves of the same coin?

Scientific enquiry is divided into two sections:

- ideas and evidence in science;
- investigative skills.

At first reading, the requirements in the programme of study may seem to suggest that these are two unconnected areas. However, many of you will have recognised that within the requirements lies the opportunity to bring together investigative science and science content. Put side by side, the frameworks underpinning each section of Sc1 look like this:

SC1 (SCIENTIFIC ENQUIRY)	
IDEAS AND EVIDENCE	**INVESTIGATIVE SKILLS**
Questions and predictions	**Planning**
	• Using scientific knowledge and understanding to turn ideas into investigations
	• Choice over source of evidence
	• Preliminary work and predictions
	• Experimental design: key factors; how evidence is collected
Evidence	**Obtaining and representing evidence** • Using equipment (safely) • Obtaining reliable evidence through observation and measurement • Representing and communicating evidence
Creative thought (and deductive reasoning) **Explanations and scientific ideas**	**Considering evidence** • Identifying patterns in evidence • Drawing conclusions; comparing with predictions • Using scientific knowledge and understanding to explain evidence and conclusions **Evaluating** • Anomalies • Sufficiency of evidence • Improvements possible

Both frameworks can be further simplified, as shown here:

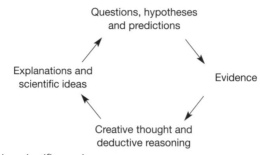

The model underpinning scientific enquiry

The current Key Stage 3 framework for Sc1 is similar to previous National Curriculum models, but with a major difference. This is the 'Ideas and evidence' dimension, which allows us to draw parallels between the investigative work pupils do and the way that scientists work and think. It also allows us to consider how currently accepted theories and models have been derived from processes not dissimilar to the ways that pupils themselves have been working in science investigations. This connection is not without problems. For instance, the model of scientific enquiry put forward in the National Curriculum is, for many, not near enough to the real world of science research. This mismatch is entirely understandable: pupils are not real scientists. We cannot expect them to come up with explanations of new phenomena in the way that scientists do. They do their investigative work largely in 50- to 70-minute bursts, often following a rigid pattern of 'plan', 'do' and 'write up' over several lessons. Whilst this is accepted, the model offers us a framework to start from.

Whilst the above model separates out the components of scientific enquiry, and places them in a sequence, it should not be seen as merely a linear progression, from hypothesis or question to evidence, to creative or deductive explanation, to a new idea, or an old one enhanced. Whilst all these processes are vital to scientific enquiry, they may in reality occur out of sequence. Creative thought may occur when working out what evidence to collect, based on a prediction about what might happen according to the application of a current scientific idea. For instance, knowledge of kinetic theory can be used to underpin the prediction that raising ambient temperature will increase the rate of evaporation from a leaf. Creative thought may be needed to identify how to measure evaporation, i.e. what evidence to collect. Deductive reasoning would also be involved in trying to develop the original hypothesis: 'Knowing what we do about kinetic theory, particles and temperature, we suggest that in the particular case of a leaf, raising the ambient temperature will increase the rate of evaporation of water from the leaf surface'. Explanation of the evidence collected will then involve a comparison with the original hypothesis. We should bear in mind this rather more fluid model of how an investigation is carried out when assessing pupils' work.

Ideas and evidence in science

School science and the public understanding of science: pupil misconceptions

We can help pupils to understand how science works. This is the main aim of the ideas and evidence section of the programme of study.

Issues like genetically modified (GM) foods, the potential risks from micro-wave radiation emitted by mobile phones, and BSE have raised concerns over the public's confidence in science and scientists. An overemphasis in previous versions of the National Curriculum for science on facts and skills has not helped. Having been brought up on a diet of factual testing and failsafe practical work it should be no surprise to us that young people think that science is merely a body of knowledge, connected in some way to a trustworthy methodology. The current requirement to teach about the nature of science, in the shape of ideas and evidence, has a role to play in educating our young people about the ways in which science works. This should allow future generations to have a better understanding of how scientific questions, experimentation and explanations are connected: in short, how scientific ideas are dependant upon reliable and valid evidence.

Pupils may believe, as they progress through their science studies at school, that science is certain about everything within its remit, i.e. the physical and biological world. Many teachers, having honed their menu of practical and investigative work to a fine art (i.e. it always works), will be inadvertently teaching their pupils that science is always like this. Where something does go wrong, it is just that – wrong. 'The experiment hasn't worked', or 'the results are wrong'. This implies that the teacher knows what should have happened but things just didn't turn out as expected, or predicted. To be fair, when school experiments go 'wrong', it is not usually due to contradictory evidence being produced which would falsify the underlying scientific idea. It may be down to errors in carrying out the procedure, or just the wrong procedure being used. It could also be due to the existence of a wide number of variables which are difficult to control, as in many biological investigations.

Almost inevitably, scientists sometimes come in for criticism. Society believes that they should have answers to important questions, such as 'Are genetically modified crops harmful to the environment?' or 'Are mobile phones safe?'. But they don't have simple answers – yes or no, for instance. Many people do not understand this – surely science is based on objective knowledge which is reliable and there to come to the aid of society when problems arise? After all, the science we learn in school is fact – we need to know the facts about the MMR vaccine. Or global warming – is it happening, what will it mean for the UK climate?

There are some areas of science which can provide us with answers to questions. Most of science is like that – science that has become accepted as being reliable. To quote Robert May, President of the Royal Society:

> *There have been advances that have hardened into virtual certitude. But the problem is that at and beyond the frontiers, that's not what science is like.*

And many of the problems that we wrestle with, and will increasingly wrestle with in future years, are problems at and beyond the frontiers of science where we don't yet know, we will sometimes have good guesses. We will sometimes have a better way of framing the question. But very often we won't have an answer.

(Professor Lord May's Science Year Lecture at the Royal Society, 2002)

THINKING ABOUT IDEAS AND EVIDENCE

- Write a list of five ideas from science which you would say are 'hardened into virtual certitude'. An example might be *the heart pumps blood around the body*.

- For each of your examples describe a piece of evidence which supports the idea.

- Write down one idea from science relevant to the Key Stage 3 programme of study which you think is still controversial, where they may be some disagreement about whether or not the idea is reliable. An example might be *living near high voltage power lines causes cancer*.

- For your example, list the evidence which supports the idea and evidence which might tend to refute it. This might include criticising the evidence in support, rather than listing evidence which refutes the idea.

- What do you think is the main reason why the idea you have suggested is controversial?

- Where does the topic fit into your scheme of work?

- What approach do you currently use to tackle the issue/idea?

- How appropriate is the approach given the nature of the issue? Is the science presented as fact or do you deal with the uncertainty?

Creative thought

A further misconception is that scientists develop ideas about science solely by applying logical thinking to the analysis and explanation of data. Star Trek has a lot to answer for! Mr Spock, perhaps embodying the scriptwriter's misconception of how a scientist thinks, literally is from another planet (and metaphorically so), as far as the rest of the crew are concerned. He is an alien species. Do pupils think scientists are like that – rejecting everything but cold logic?

We need to make pupils aware of the rich diversity of thought that results in scientific ideas and explanations. Yes, a logical approach is at the heart of the scientific enterprise, but there is imagination and creativity as well. Examples from history (such as the 'snakes chasing their tails' dream which gave Kekule the idea for the structure of the benzene ring) or from more recent times (such as Harry Kroto and the discovery of buckyballs) show how logical thought alone was inadequate. Neither chemical structure would have come to light without the creativity of scientists.

Pupils can be introduced to the use of creative thought and imagination in science by dealing with models of phenomena which they cannot see directly, such as sub-atomic particles, magnetic fields and heat transfer. We can make it more explicit that when thinking of phenomena and explanations like this, pupils are having to use their imagination, and that the original idea behind the particular model or explanation is a product of human creativity. Somebody had to come up with the idea which seemed to fit the available data. Explanation, a creative act, is not 'out there' (i.e. waiting for someone to discover it) but 'in here', inside someone's head. It is then shared with others, first scientists and eventually the wider public.

What are we trying to teach about the nature of science?

For many young people, their interaction with scientific explanations is mainly through school. This is not to be confused with their interaction with science, which can come from their everyday lives, as consumers, users, viewers, listeners, participants and observers. This is particularly so in terms of their interaction with the technological products of science, although the media also present them with many science-based stories. How does school science connect with their everyday experience of science? A recent report (Osborne and Collins, 2000) highlighted the problem here: *pupils do not see the connection.* To these youngsters, school science seems more concerned with getting through the syllabus – the overloaded curriculum often based on historical examples and where applications are dealt with as an afterthought, or so it seems to the pupils. There is little opportunity to deal with issues, to look at how a theory came about, to consider the relationship between an explanation and the evidence which may, or may not, support it. In short, if it were not bad enough that pupils feel let down by a subject which has the

CASE STUDY – STEPHEN SPARKES AND VULCANOLOGY

One of the requirements of 'ideas and evidence in science' is that pupils should be taught about *the interplay between empirical questions, evidence and scientific explanations using historical and contemporary examples.* Stephen Sparkes is a contemporary scientist working in the UK. His work has helped change vulcanology from being a predominantly observational science to one which uses experiment and investigation much more.

Sparkes' work involves looking at the link between how fast material is ejected from a volcano and the type of volcanic eruption. One of his hypotheses, which grew from attempting to explain his and others' observations of eruptions, suggests that air rushing into the hot ash during an eruption was a significant factor in determining the speed of flow of the ash (pyroclastic flow). This gave rise to empirical questions, involving the testing of the hypothesis and the production of experimental evidence. In this case, the evidence collected supported the original hypothesis. For further information on Stephen Sparkes' work see the Acclaim Project materials (Centre for Science Education, 2001).

ability to inform so much of their everyday experience, what they do learn is considered to be not just irrelevant but also virtually devoid of human input. Science is what is in the text book. How it got there is not just unknown, but irrelevant. School science seems to be the process of getting young people to learn all this stuff, rather than dealing with how we know it all in the first place. It seems even more of a long shot to aim to teach youngsters to appreciate what we mean by 'know' anyway!

This way of looking at how Stephen Sparkes works is useful in drawing parallels between 'Ideas and evidence' and 'Investigative skills'. Pupils learn that questions, hypotheses or predictions (or a combination) are the starting points for investigations. They then collect evidence and consider what this means in terms of their starting point. Does it answer the question? Is the hypothesis supported? Was the prediction correct? The use of contemporary or historical examples of real scientists' work, framed as in the above example, can help pupils make the link between their own investigative work and how scientists (and science) work. Also, by showing how certain science concepts came to be accepted, this approach can also provide a bridge between 'Scientific enquiry' and the knowledge and understanding elsewhere in the programme of study.

CASE STUDY — SIR PAUL NURSE AND CANCER RESEARCH

One of the requirements of 'Ideas and evidence' is that pupils should be taught *that it is important to test explanations by using them to make predictions and by seeing if evidence matches the predictions.*

In 2001 Sir Paul Nurse, a British scientist, won the Nobel Prize for medicine, following his work on the genetic control of cell division. Uncontrolled cell division is one of the features of cancer.

Nurse was working with yeast cells. He came to understand how a gene in yeast could control cell division in the yeast. Using his knowledge of how the gene worked, he predicted that similar genes might occur in other species, and that all living things might control cell division in the same way. His team of scientists carried out experiments, including trying out a range of human genes as replacements for the cell division gene in yeast (in which the cell division gene had been inactivated). They found that one human gene actually made the yeast cell divide. This obviously implied that the way in which the cell cycle was controlled is basically the same in all living things, from yeast to human cells.

Paul Nurse made his prediction, and then carried out investigations which showed that his original prediction was correct. His work has led to a greater understanding of how cancer occurs.

Let us look at a second example – the work of Sir Paul Nurse, head of the Imperial Cancer Research Fund (now part of Cancer Research UK). In this example it is easy to identify the prediction, the background or underpinning knowledge, and how investigative work provided evidence that the prediction was correct.

INVESTIGATION AND SCIENTIFIC KNOWLEDGE

Select one of the following examples of common laboratory investigations and identify the factors required to complete the chart.

Common investigations:

- fermentation of yeast: effect of temperature;
- effectiveness of different antacids;
- cooling rate and size of container.

Identify the following.

INVESTIGATION	
Underpinning scientific knowledge and understanding	
Prediction	
What evidence to collect	

Now complete the same sort of table for one of the following:

- the development of pasteurisation as a technique to stop milk going sour;
- the use of 'lime' in water treatment;
- the design of an aerodynamic wing.

The aim in the above activity is not to think about doing the investigation, but to identify the components (background knowledge, prediction, evidence) which allow us to see how investigation and scientific knowledge link together in a more explicit way than is usually presented to pupils.

Investigative skills

Investigations can be very motivating for pupils. They can also be a turn-off. It all depends on how you do them. Giving pupils opportunities to develop their skills through a variety of activities across the entire key stage, with occasional whole investigations, is better than leaving it all to a small number of 'assessed investigations' towards the end of Year 9.

Scientific enquiry is developing slowly. In most schools, pupils undertake just two or three whole investigations a year often to a closely prescribed pattern. There is growing separation between practical activity for coursework assessment and its use as part of normal teaching.

(OFSTED, 2002)

We begin by looking at some of the problems and issues raised by the need to teach investigative skills, and then go on to look at some of the possible solutions.

Investigations and practical work

Not all investigations involve practical work, and not all practical work is an investigation. This may be obvious, but it gives us a good starting point.

Why do we do practical work in science? What are our aims? Many lists have been produced, but this is perhaps the most succinct.

- To aid pupil motivation.
- To develop pupil skills.
- To assist in the learning of scientific knowledge.
- To teach the methods of science.
- To develop scientific attitudes.

(Hodson, 1993, p 90)

Although motivation is seen as a key aim, we must remember that whilst many pupils enjoy practical work, others do not. Practical work may be seen as a break from the usual teacher-led lesson, but there are other ways to provide a more varied diet of learning experiences than just 'doing a practical'.

The other aims in the list can be summarised as: to aid the development of scientific knowledge, and to teach about how scientific enquiries are carried out.

Obviously, pupil skill development is important. OFSTED has repeatedly commented on the lack of a coherent and planned approach to developing pupil skills.

Perhaps the most difficult aim is to use practical work to develop pupils' knowledge of science concepts. Evidence shows that a high degree of teacher intervention is required here. Pupils do not find it easy to move from observation or measurement to explanation, and from here to a new understanding of a concept. In fact, the link between observation (a steel bar picks up paper clips when a current is passed around it through a wire) to explanation and theory (electromagnetism) is often impossible for pupils to make. This is where the teacher fits in, with an explanation. Much practical work fits this pattern. In general, such activities are 'safe' for the teacher – they are predictable and hold few surprises. They provide pupils with the experience of producing hydrogen from acid/metal mixtures, or of showing that photosynthesis has occurred in a geranium leaf. Explaining the observations (bubbles of gas; black stain with iodine) requires use of accepted scientific theory. These activities allow teachers to provide pupils with the experience of seeing the bubbles, so that when they learn about the reaction of (some) metals with (some) acids they have a feel for the phenomenon. But the activity itself does not 'teach' about displacement or the reactivity series as such.

Using a data logger to record changes and plot these as a graph in 'real time' allows pupils more opportunity to observe the changes taking place. Often pupils' attention is so focused on measuring and recording that they can miss the purpose of the activity, such as the change of state (when salol cools) or the colour change (titrations, enzyme reactions). The beauty of using a data logger is that pupils can relate the shape of the graph to the events taking place before them as it actually happens.

Digital projectors and interactive whiteboards are useful tools for drawing together different points from a lesson. They are particularly effective at helping you to aggregate the results from group experiments, and to help pupils make links between observations and explanations. Using changes of state as an example, a computer simulation of particles in solids and liquids may be used to help explain the shape of graphs obtained when ice melts.

Teaching about the methods of science is also problematic. There is no single methodology which takes in the variety of approaches used in science. Scientific method cannot be seen as a single hierarchical set of processes. It is more useful to look at the 'procedural' understanding which we try to develop in pupils in a similar way to the development of conceptual understanding.

Types of practical work

From the myriad of different practical activities which go on in science classrooms it is possible to identify at least three broad types, which address different aims.

TYPE	AIMS
Illustrative practical work	Conceptual development, particularly subject knowledge and understanding. Learning science.
Investigations	Procedural development. Learning about science and doing science.
Skill-developing practical work	Skill development. Doing science.

In actual fact, as discussed above, illustrative practical work is not very effective in developing pupils' knowledge and understanding of science. It does provide a set of experiences and skills which we think pupils should have and develop, such as using apparatus safely and making measurements and observations. Pupils in most schools will observe a range of chemical reactions, such as the action of acid on metals and on carbonates. They will also detect starch production in leaves and make simple electromagnets. Teacher intervention is required to help the pupils make sense of the results they obtain – although they may be able to rank metals into reactivity series order, they will not leap from here to a new understanding of the periodic table. Whilst this sort of practical work can be used as a starting point for developing conceptual understanding, are we missing a trick here? Are we using such practical activities to develop pupils' investigative skills?

USING ILLUSTRATIVE PRACTICAL WORK TO DEVELOP INVESTIGATIVE SKILLS

When planning for practical work, rather than produce a set of instructions for pupils to follow, or a skillful demonstration, aim to begin the session by asking pupils key questions about the experimental design.

- Begin with the underpinning scientific concepts.
- Identify a key question which might arise from thinking about this creatively.
- Lead pupils into making a prediction based on the question.
- Depending on the type of investigation being undertaken, identify the key factors which might be involved.
- Deal with the factors, by allocating them to one of the following categories: factors which need to be controlled (control variables); a factor which you will systematically vary (independent variable); the factor which you will observe or measure as you change the independent variable.
- Talk about the range of values you will use in manipulating the dependent variable, and why a particular range is more useful than a wider or more narrow one. This is where pupils' understanding of the background science is important.
- Explore how the experiment might be carried out – the apparatus needed, how to control the variables, how to collect the evidence (observations or measurements).
- If the pupils are carrying out the investigation, they now have an understanding of both the purpose of the activity, and how the procedure has been designed.
- If you are doing a demonstration, the involvement of the pupils in the design will increase their involvement as they observe what is going on.

TEACHING INVESTIGATE SKILLS

- List the last five pieces of experimental or investigative work that you carried out in school.
- Select one which you do not normally use to teach investigative skills.
- Follow the sequence in the above box and identify ways in which you could introduce the activity so as to emphasise investigative skills.
- Repeat this activity with other examples of non-investigative practical work in order to increase the opportunities for teaching about investigative skills.

Ideas for the use of ICT to support the development of investigative skills

DATA LOGGERS

Some data logging software allows the graph line to be temporarily turned off and for a predicted line to be added. Pupils cannot always explain their predictions, but enjoy seeing who can get closest to the actual result.

Data loggers can reveal unexpected results – try comparing the cooling effect of evaporating water with that of an alcohol. Questions that arise from analysing such graphs can lead to further investigations being carried out in the same lesson. How many repeated readings are needed to obtain a reliable average? Using light gates to measure the speed of a falling object is a quick way to teach this important principle, and the software takes care of the mathematics. This could be part of a lesson involving measuring the speed of a toy car from different positions on an inclined dinner tray (a dinner tray stops the car falling on the floor). How many positions are needed to obtain a reliable pattern? Is speed proportional to the distance up the incline? Is the same pattern true for other angles of inclination? Data loggers enable pupils to answer these questions effectively in a short space of time.

COMPUTER SIMULATIONS OF PRACTICAL WORK

Simulations of investigations are useful additions to departmental resources. They allow pupils some experience of hazardous, expensive or problematic activities for themselves. They are very useful for helping pupils to identify variables and select appropriate ranges of values for investigation. The example of measuring the bounce of a ball is well documented in Key Stage 4 science investigation literature, but other examples include industrial processes, radioactive penetration and rates of photosynthesis. At Key Stage 3 some of these simulations are appropriate for teaching pupils how to develop an increasingly quantitative approach to investigations.

SPREADSHEETS

Spreadsheets are good for aggregating class results when used with a digital projector. This could then lead on to calculating averages and identifying anomalous results. Some teachers initially find the graphing tools cumbersome and unpredictable, but it is well worth investing some effort in mastering them. Spreadsheet programs like *Excel* contain some powerful data analysis tools, probably better suited to older pupils, including the ability to add a line of best fit (trendline) and the equation for the graph.

Types of practical work

The AKSIS (ASE and Kings College Science Investigations in Schools) project produced a useful classification of types of practical work found at Key Stages 2 and 3. The main types they identified are:

- classification and identifying;
- fair testing;
- pattern seeking;
- exploring;
- investigating models;
- making things or developing systems.

The AKSIS research at Key Stage 3 showed that the vast majority of investigations (85 per cent) were fair tests. OFSTED has frequently commented that the range of investigations in school science is too narrow. It is not surprising that many teachers develop and then stick with a small number of highly predictable, safe, controllable practical activities. These are the assessed practicals which allow pupils to score the highest marks. There are two problems with this. Firstly pupils are not being introduced to the wider range of types of scientific activity, so their view of science – of what science is, how it works, what it can do – is biased towards the 'right answer' model. In the age of BSE, cloning and mobile phones, scientists are coming under increasing criticism because of the tentative nature of the knowledge they are communicating to the public, who are waiting for definitive answers. The safe and predictable science practical reinforces the wrong view of the nature of science.

Secondly, pupils are not developing the wider range of investigative skills which depend on experiencing a more varied diet of investigations. Nor are many of them being taught even the 'core skills' required for fair test investigations in a planned and systematic way. However, it may be even worse. According to OFSTED reports, pupils in Key Stage 2 are doing whole investigations, albeit mainly of the 'fair test' type, but these are usually more open-ended and exploratory than those used in Key Stage 3. We are not building on these skills following transfer to secondary school.

> *Practical work designed to demonstrate principles in Key Stage 3 often displaces the more open practical activity that pupils have been used to in Key Stage 2. This means that they experience an abrupt transition and are consequently not able to demonstrate fully the skills and understanding gained earlier.*
>
> (OFSTED, 2000)

Widening the range of investigation types

The following table uses the six categories identified by the AKSIS project as a starting point.

CATEGORY	DESCRIPTION	EXAMPLES
Fair test	Investigating relationships between variables or factors. Usually one factor is systematically varied, (independent variable) and its effect on another factor (dependent variable) is measured or observed. Other variables are controlled.	Effect of temperature on rate of reaction.
Pattern seeking	Approach used where it is difficult to control all variables or factors such as in living things and ecosystems. Involves making measurements and/or observations of a factor, such as plant distribution, and seeking patterns and correlations with possible causal factors.	Factors affecting distribution of plant species in specific habitat, e.g. hedgerow, pond area, meadow.
Classifying and identifying	Identifying organisms, materials, objects. Grouping these according to criteria.	Identification of animal and plant species in habitat.
Exploring	Observation of objects or phenomena, including changes over time.	Observation of rock types, development of tadpole to frog.
Making things or developing systems	Technological tasks with a significant scientific content.	How can an electric motor be made more efficient?
Investigating models	Testing a theoretical model using one of the above approaches.	Investigating insulation and cooling using models of heat transfer.

When planning a scheme of work, it is important to try to achieve a balance across these types. That does not mean equal numbers – fair tests are likely to be the predominant type. However, it should be possible to increase the number of examples in the other types so that the menu experienced by pupils is not so one-sided.

LOOKING AT INVESTIGATION TYPES

- List all the investigations you carry out at Key Stage 3.
- Allocate each activity to one of the AKSIS categories.
- What proportion of each type do your pupils experience?
- How can you increase the number of investigations in categories in which you have few (or no) examples?

Developing pupils' investigative skills throughout the key stage

Pupils will begin Key Stage 3 with a wide variety of experiences of doing scientific investigations, including carrying out whole investigations, either on their own or in groups. At Key Stage 2 the types of investigation carried out are often of a more open nature to those at Key Stage 3, and it is sometimes a culture shock for pupils to experience the more closed or constrained approach at Key Stage 3.

The *Framework for Teaching Science Years 7, 8 and 9* (DfES, 2002 (a)) sets out a series of yearly teaching objectives for Sc1 (scientific enquiry). These describe what pupils should be taught in each of the three years. It provides a useful checklist when planning a teaching programme for this aspect of science. The Year 7 objectives build on the learning pupils will bring with them from Year 6. The Framework identifies for teachers the progression into and across the key stage.

The Framework shows how pupils' procedural understanding can be developed. However, it is also important to consider the conceptual demands of the underpinning scientific knowledge in each enquiry. Evidence shows that progression in learning is best achieved by being aware of the balance between the two sets of demands. Clarity of objectives is also important: is the focus for an activity the development of scientific knowledge or the skills of scientific enquiry?

CONCEPTUAL UNDERSTANDING	Demands	PROCEDURAL UNDERSTANDING	
		Low	High
	Low	1. Task is based on less demanding scientific knowledge and requires low level of procedural understanding. May consolidate but not develop pupil understanding of either area.	2. Task is based on scientific knowledge with which pupils are reasonably secure but procedures are demanding. Useful in developing understanding of scientific enquiry.
	High	3. Task is based on demanding scientific knowledge but procedure is already understood by pupils. Useful in developing scientific knowledge.	4. Demands of both scientific knowledge and procedures are high and pupils may not be successful in developing both aspects through same activity. However, such activities can be useful in providing challenges for pupils.

PROCEDURAL AND CONCEPTUAL DEMANDS

- Use the list of investigations you made for the previous activity.

- Assess the procedural and conceptual demands of each investigation.

- Allocate each investigation to one of the four numbered quadrants from the above table. Quadrant 2 would seem the best for developing pupils' understanding of the procedures important in pursuing a scientific enquiry.

- For those in quadrant 1: how could you increase the procedural demands of this investigation for pupils who would not find this a challenge?

- For those in quadrant 3: assess the aim for this activity. If it is to develop procedural understanding, identify which skills you are seeking to develop. How can pupils be supported in these areas, given that the conceptual demands of the activity are high?

- For those in quadrant 4: assess the aim of this activity. With both high conceptual and procedural demands your pupils may find this activity difficult. Identify which skills you are seeking to develop. How can pupils be supported in these areas, given that the conceptual demands of the activity are high? How can the conceptual demands be supported? Is this activity used with pupils who find investigations difficult?

The following table shows how one strand within the requirements for Scientific enquiry might progress from Years 7 to 9.

YEAR GROUP	OBJECTIVE	NOTE
Year 7	Deciding approach to how an idea is tested	Pupils are likely to have experience of deciding approaches at Key Stage 2. Consolidation of pupils' understanding can be first achieved through use of more demanding concepts at Key Stage 3.
Year 8	Making judgements about which approach from several might yield better evidence	This requires knowledge that there may be more than one way of tackling an investigation, and sufficient experience to judge which approach might produce better evidence. Development towards this objective may initially require using scientific knowledge familiar to pupils in contexts where they try out more than one approach to gathering evidence.
Year 9	Making judgements about approaches that are suitable for different types of question	This requires understanding that different approaches to an investigation are appropriate for different types of question (see AKSIS categories). Development towards this objective may require pupils to experience a wide range of investigation types, and to understand the circumstances in which each approach is appropriate.

Other strands of progression across Key Stage 3 identified in the Framework relate to:

- ideas, evidence and explanation;
- identification and control of key factors;
- selection and use of apparatus;
- achieving reliability;
- presentation of results and interpretations;
- explanations and conclusions;
- evaluating evidence.

The development of pupils' understanding of these areas can be planned across the key stage – the Framework document is useful here in identifying the yearly teaching objectives.

What is then essential is to identify opportunities in the scheme of work for the development of the underlying procedural concepts, bearing in mind the interplay between conceptual and procedural demand. Put simply, when developing a new procedural concept, it should be through a context that uses scientific knowledge in which the pupil is secure. Difficult content should not get in the way of developing procedural understanding.

Progression in scientific enquiry and the QCA scheme of work

It is not clear to what extent the issue of planning for progression in the concepts described in 'Scientific enquiry' is evident in the guidance for the development of the concepts as set out in the QCA scheme of work.

If the scheme is followed in unit order, then Unit 7A Cells will be the first one studied in Year 7. The first example of a scientific enquiry is the observation of cells using a microscope. Later there is an investigation into the optimum sugar concentration for pollen tube growth, involving the concept of sample size.

The microscopy activities are used to introduce pupils to how evidence from microscope observations changed ideas about the structure of living things. This includes a look at how our first ideas about cells were dependent on the development of simple microscopes, and how our understanding has grown over the centuries as better instruments have been produced. This is also an ideal time to discuss the place of observation in scientific enquiry, and the fact that some enquiries are based on the technique.

The place of the pollen tube investigation should be questioned. It involves pupils in dealing with high demand scientific knowledge – the role of sugar in growth, and growth itself, as well as high level procedural knowledge, including new techniques and the idea of sample sizes.

Opportunities for developing pupils' understanding of scientific enquiry

Departments should carry out an audit of the practical and other investigative work they carry out in Key Stage 3. The output from some of the activities earlier in this chapter will be useful here. The audit should involve all practical tasks, even those which have not so far been intended to contribute to developing the skills of scientific enquiry, such as demonstrations and worksheet based practicals.

AUDIT OF PRACTICAL WORK	
ISSUE	**TARGET**
The range of investigation types pupils meet when carrying out (or observing) the tasks	Departments should aim to have a balance of investigation types.
The procedural and conceptual demands of each task	Aim not to have too many Quadrant 1 and 4. Quadrant 2 is ideal for developing pupils' knowledge of scientific enquiry. Quadrant 3 is more suited for work where the key objective relates to pupils' understanding of science concepts.
What aspects of scientific enquiry are developed through the task	Using the yearly teaching objectives from the Framework, identify which aspects of scientific enquiry are relevant to the task. Identify the sequence in which pupils use these aspects – is progression evident? Are they dealing with high level demands before lower level demands?
What is the balance between whole investigations and tasks dealing with only one or two aspects (if any) of scientific enquiry?	Departments should have a range of whole investigations across the key stage, not just as assessment opportunities during Year 9.
How is scientific enquiry integrated into the teaching of other science content? Particularly how is 'Ideas and evidence' integrated into the teaching of other science content, and the 'Investigative skills' section of 'Scientific enquiry'?	Although set out separately in the programme of study, 'Ideas and evidence' and 'Investigative skills' share a common model for the link between ideas (and explanations) and evidence. Is the concept of evidence when relevant to teaching science knowledge (e.g. how scientists relate some diseases to a lack of certain nutrients) reinforced when pupils consider the evidence they acquire in a practical investigation?

An audit which looks at these questions is likely to uncover the following issues:

- There is a preponderance of 'fair testing' investigations.
- Much practical work is not seen as contributing to the development of scientific enquiry.
- Little account is taken of balancing the competing procedural and conceptual demands of a task to encourage pupil learning.
- There is little coherence in how pupils are taught aspects of scientific enquiry across the key stage.
- Whole investigations are mainly used for assessment purposes.
- Aspects of 'Scientific enquiry' which deal with 'Ideas and evidence' are not currently integrated into the scheme of work, and this is possibly the case

PRODUCING AN ACTION PLAN

Depending on the degree to which the above points match your analysis, the next job is to set out an action plan for ensuring that each issue is addressed. An action plan is likely to contain elements such as the following.

- Increase the number of tasks which are:
  - classification and identifying;
  - pattern seeking;
  - exploring;
  - investigating models;
  - making things or developing systems.

- Identify how tasks (including demonstrations) not currently contributing to 'Scientific enquiry' can do so. For example, by involving pupils in decision making about experimental design, choice of measuring instruments, range of measurements, etc.

- Identify the demands of the required scientific knowledge of each task, and of the required procedural understanding. Having established the purpose (objective) of each task, adjust the demands to allow the objective to be reached more effectively.

- Make a plan to show how aspects of 'Scientific enquiry' are currently sequenced in the scheme of work. Re-plan to provide for better progression across the strands.

- Identify opportunities to insert more whole investigations into the scheme, particularly at the end of a unit.

- Develop an assessment framework which allows:
  - pupil target setting;
  - on-going collection of assessment evidence;
  - periodic assessment using whole investigations.

- Identify where there are opportunities to teach about 'ideas and evidence' in the scheme of work, and make links with how pupils consider evidence gathered through their other investigative work.

both for integration with other subject content, and with the 'Investigative skills' component.

Summary

In this chapter we have looked at how we can teach pupils about science, about doing science, as well as how we teach them science (although this last point is dealt with more fully in Chapter 3). These may all seem to be the same thing, but they are not.

Teaching about science is concerned with looking at how science works, what is sometimes referred to as the nature of science. Teaching about doing science focuses on investigative skills, in particular those involved in doing certain types of practical work, but there is more to it than that. Teaching science usually relates to dealing with the facts and concepts of science.

Sc1 (Scientific enquiry) allows us to bring these three areas together, although we must always be clear about the specific learning objectives of any activity. In particular, we have noted the integrating role of ideas and evidence in science, and how this can provide a bridge between investigative work and science content. 'Ideas and evidence' allows us to view investigative work and science concepts through the same lens.

We have looked at broadening the range of types of investigations we use at Key Stage 3 and how to increase the opportunities for developing investigative skills by adapting existing practical activities.

In all, we have laid out some of the issues related to placing Sc1 (Scientific enquiry) at the heart of teaching science, rather than seeing it as an add-on dealt with through occasional practical assessments. Not only is this a positive move in terms of pupils' experience of school science, it is also positive in terms of the public's appreciation of the nature of science.

In summary, this chapter has highlighted the need to:

- identify aspects of the Key Stage 3 curriculum that are based on science which is still uncertain, for example the causes of global warming;
- identify how investigative work and scientific knowledge link together in a more explicit way than is usually presented to pupils;
- plan so that the investigative work pupils meet in science is balanced across the range of investigation types;
- use a wider range of practical activities to develop pupils' investigative skills;
- plan for the development of pupils' investigative skills through considering the procedural and conceptual demands of investigations;
- Encourage the development of pupils' investigative skills by producing a Key Stage 3 plan which uses the end-of-year objectives from the Framework.

Further reading

Wellington (ed) (1998) *Practical work in school science: which way now?* London. Routledge.

An excellent book looking at all aspects of practical work, from a considered, theoretical perspective. Challenges many suppositions about the role of practical work in science.

Centre for Science Education, (2001) *The Acclaim project.*

A curriculum pack aimed at raising young people's awareness of the work and lives of eminent living scientists and engineers. Very useful for *Ideas and evidence* work.

Science content: key ideas at Key Stage 3

This chapter focuses on what science we teach. The main emphasis is on the key ideas which underpin the development of pupils' understanding.

Pupils' attitudes towards science at the start of Key Stage 3 are generally positive and teachers know that this is maintained best by relating science to their pupils' needs and interests. This positive attitude can be enhanced if we ensure that pupils not only make progress in their understanding of scientific ideas but also learn to appreciate the significance of science in their lives.

Research suggests that during the course of Key Stage 3 pupils' attitudes towards science can decline and many lower-attaining pupils make little progress. It seems that in particular they find it difficult to make the step from the acquisition of information to an understanding of the concepts that are so important in science. This chapter will examine the issue of the key ideas underpinning teaching and learning at Key Stage 3 and look at some possible strategies that could be employed by the classroom teacher to help deliver them.

The topics to be covered include:

- key ideas in science at Key Stage 3;
- contemporary science;
- teaching about controversial issues;
- using contemporary science and issues as starter activities.

KEY ISSUES

In this chapter, you will:

- understand how the key concepts, theories and models can be used in a range of topics and contexts so that you can encourage your pupils to make connections between these topics;

- be able to plan lessons which deal with controversial and/or ethical issues in a sensitive way;

- appreciate the speed of change in knowledge and understanding in certain areas of science;

- understand the importance of choosing strategies that build on your pupils' interests and experiences but also be aware of the value of using contemporary applications in teaching.

Key ideas in science at Key Stage 3

Relevance of Key Stage 3 programmes of study to key ideas

The following is an outline of the importance of identifying the major key ideas within the programmes of study at Key Stage 3, which most of you will recognise.

- To progress in science your pupils must not only develop a knowledge of the key scientific ideas, but they must also understand how these underpin the development of their wider scientific knowledge.

This raises an issue you will almost certainly be aware of:

- This type of teaching and learning is not just a case of memorising more facts. To be effective it requires pupils to understand these scientific concepts and, at the same time to develop the ability to apply that understanding to new and different situations.

What does this mean to you, the classroom teacher?

- It is a major challenge at Key Stage 3. Pupils will need to develop and use abstract ideas and their application and use models and analogies to represent them.

Recent inspection findings across Key Stage 3

Recent school inspections of Key Stage 3 science have shown that the levels of attainment at the start of Year 7 are higher than in previous inspections. This is thought to be due to the improvements in primary science. In addition, compared with earlier inspections, pupils in Year 7 showed improvements in their understanding of key scientific ideas such as the particulate nature of matter.

By the end of Year 8, pupils' knowledge seemed to have been extended and applied in a wider range of contexts. For example, the idea of particles was being used to describe and represent chemical reactions.

You may have noticed an improvement in progress in Year 9. Inspectors found that this was particularly true for the higher attaining pupils. However, the lower attaining pupils seem to make less progress. This often seems to be because they have failed to make the important leap from simply acquiring more information to developing conceptual understanding.

What is a 'key idea' in science?

A key idea is a prerequisite for understanding other concepts. It underpins learning across a wide range of scientific ideas and concepts and can be applied when explaining a range of phenomena.

At Key Stage 3, there are five key ideas:

1. Cells
2. Interdependence
3. Particles
4. Forces
5. Energy

Teaching key ideas

Why teach key ideas and what are the implications for teachers?

Key ideas:

- are needed for pupils to attain higher levels attainment at Key Stage 3;
- need to be understood by teachers first if these higher levels are then to be achieved by pupils;
- highlight the importance of the order of teaching.

For pupils to develop their understanding it is important that these key ideas are introduced early in Key Stage 3. That way pupils will not only recognise them as important but will gradually learn how to use them and eventually how to apply them in different contexts.

Raising standards – how can teaching key ideas help?

If you want to raise standards you should:

- emphasise the importance of key ideas in science to help pupils clarify their prior thoughts and ideas about a scientific concept – this may often involve dealing with pupils' misconceptions;
- enable pupils to develop their understanding of key scientific concepts;
- encourage pupils to apply this new learning to a range of subjects.

When planning for teaching and learning science remember that scientific language is technical and care must be taken when introducing new terms. However, this is an opportunity for you to draw your pupils' attention to the use of words and terms across all three content areas of the programme of study, and, just as importantly, how these words and terms are interrelated. Scientific ideas can be abstract and can be applied in different contexts. You will need to give your pupils practice. The best way to start is with everyday examples and they will learn the process quickly.

It is a useful strategy to identify and collect examples from newspapers or news reports on television which deal with science related to the key ideas – for example, energy conservation issues and the use of fossil fuels. By reading/watching them, your pupils can be encouraged to discuss the science content: Is it correct? Is it explained sufficiently well for the average person to understand? Can they now make informed decisions about the issue being dealt with? Could they (the pupils) do better at writing a similar article?

They can then devise their own newspaper or television report and deliver it to the class. This approach should also help you to develop your skill of dealing with questions from pupils about issues of a potentially sensitive and/or controversial nature.

Key ideas within and across the Key Stage 3 programmes of study

Some key ideas are specific to different parts of the programme of study while others span all three content areas. The key ideas that can be identified across Sc2, Sc3 and Sc4 can be considered to be fundamental to an understanding of science, and it will be no surprise that they are 'Particle theory', and 'Energy transfer and conservation' as seen in the following table:

SC2	SC3	SC4
Cells as the building blocks of life		
Particle theory	Particle theory	Particle theory
Energy transfer and conservation	Energy transfer and conservation	Energy transfer and conservation
Interdependence		

Why are these key ideas?

CELLS

By ensuring that your pupils understand what cells do and how they can become specialised, you are helping them understand the complexity of living organisms.

INTERDEPENDENCE

Pupils need to appreciate that there is a fundamental interdependence of the biological and physical environments and that the Earth is an inter-related environment and not a group of distinct parts.

PARTICLES

This is an abstract idea but it underpins *all* science study at Key Stage 3. If your pupils understand the particle theory of matter, they will start to understand and be able to explain the nature of materials and how they behave.

FORCES

At Key Stage 3 pupils need to develop a more abstract view of forces, such as an understanding of how forces can balance out, although more detail is not needed until Key Stage 4.

ENERGY

Energy is another underlying abstract idea that will help pupils explain a range of phenomena. By using examples and models you can demonstrate the concept of energy conservation, although the detail is not needed until Key Stage 4.

All teachers will appreciate that there is a close relationship between the three content areas of the programme of study (Sc2, 3 and 4) and how concepts in one may be underpinned by key ideas from the others.

There are a number of examples of these interrelationships and as a classroom teacher it is a worthwhile exercise to identify as many as you can so that you can take account of them when you are teaching. To give some starting points, three examples are given below.

The Sc3 and Sc4 key idea of particles underpins concepts found in Sc2 such as:

● cells and cell function – pupils need to understand the difference between cells and particles;
● respiration – the reaction between oxygen and food (word equation);
● photosynthesis – reaction and equation (word equation).

The Sc4 key idea of energy underpins concepts found in Sc3 such as:

● particle theory – to explain solids, liquids and gases, change of state, gas pressure, diffusion;
● solubility – its variation with temperature;
● chemical reactions which give out energy.

The Sc3 key idea of particles underpins concepts found in Sc4 such as:

● forces – air resistance (pupils need to understand that air is a gas);
● hearing how sound travels and that it cannot travel through a vacuum;
● conservation of energy – the distinction between energy and heat and that differences in temperature can lead to energy transfer;
● conduction, convection and evaporation.

The underpinning key ideas across all parts of the programme of study have been identified as particle theory and energy. Just where can they be found in the National Curriculum at Key Stage 3? The answer is: in lots of places, but they are not always obvious and it is easy to forget that an understanding of them may be needed for pupils to progress their knowledge and understanding of other phenomena.

For example:

IN Sc2

Understanding of cells and cellular processes needs to be underpinned by an appreciation of particles and energy if they are to be fully understood, e.g.

53

cellular processes in respiration and photosynthesis, the chemical reactions taking place and the role played by energy in each one.

IN Sc3

Energy transfer and particle theory are frequently linked together to explain phenomena, e.g. weathering of rocks and the energy required for particle transportation.

IN Sc4

Energy transfer and particle theory are frequently linked together to explain phenomena, e.g. sound travelling through a medium requires the movement of particles.

What to teach and when

It is important to consider the teaching order of topics that will be relied on later as key ideas. This is especially true in Year 7 where foundations are often secured.

For example, which would you teach first?

- Solids, liquids and gases or the rock cycle?
- Solids, liquids and gases or sound?
- Energy transfer or circuits?
- Energy transfer or change of state?

The five key ideas need to be introduced early in Key Stage 3 so that they can be used as a foundation for later work. Look at the following scenarios. Given the interrelationships mentioned earlier, what would you do?

CONSIDERING TEACHING ORDER OF CONCEPTS

- Using the QCA Key Stage 3 Scheme of Work, construct two concept maps, one around particle theory and one around energy transfer and conservation, showing the interrelationships between the topics and Sc2, Sc3 and Sc4.
- What are you going to teach first and why?
- Now come up with a 'running order'.

This is another opportunity to show your pupils the connections between two fundamental scientific concepts. But in this case, the interrelationships are ones that span Sc2 to Sc4 and can therefore be seen as fundamental to the study of science at Key Stage 3.

Misconceptions and simulations

Computer simulations are an effective way of getting pupils to discuss their ideas about difficult concepts. Animations of changes of state help pupils to understand that the particles themselves do not change state, but that their proximity and motion are influenced by the amount of energy available. The interactivity of good software encourages pupils to explore their understanding, and of course, computers make no value judgments when unusual responses are made. Simulations are usually based on a simplified model of the science concept. They provide a readily accessible view, and consequently offer opportunities for you to explore the limitations of the model with the class. If you do this as a class activity using a digital projector you have a ready means of assessing the understanding of your pupils.

Progression

The key ideas provide a framework that could help reduce the fragmentation of the curriculum which could occur if units were taught with little reference to sequencing and cross linking. Progression in pupils understanding of scientific concepts needs careful planning, but if it is done effectively your pupils will gain confidence and the ability to relate key ideas across the whole programme of study. Your pupils should be building on knowledge from Key Stage 2 and, at the same time, adding to their knowledge with new topics and ideas. You should plan and have an expectation that they will make connections across the programmes of study and maybe even beyond.

PLANNING FOR DEVELOPMENT OF UNDERSTANDING

By examining how each of the key ideas progresses from Key Stage 2 to Key Stage 3, you can plan for the development of pupils' understanding.

- Look at the yearly teaching objectives in the *Framework for Teaching Science*. These show the way that each of the key ideas can be developed throughout Key Stage 3, and some of the ways in which each idea can be used to underpin learning in other units.

- For each key idea, chart its progress and development from Year 6 to Year 9. Show the units (from the QCA Scheme of Work or your own unit titles) in the order in which they should be met by pupils so as to strengthen their understanding of the key ideas and other ideas which depend on this understanding.

This should help you develop a range of teaching and learning approaches that will enable you to deal with pupils' misconceptions and their everyday thinking about science. It will also give you the opportunity to introduce some contemporary applications in your teaching.

PROGRESSION MAPPING	UNITS IN Y6	UNITS IN Y7	UNITS IN Y8	UNITS IN Y9
Cells				
Interdependence				
Particles				
Forces				
Energy				

Use this information when planning the details of the scheme of work for your department. One important aspect of this activity is the need to be aware of what your pupils were taught in Year 6. Knowing this will help you to reduce unnecessary duplication of work.

Subject specialisms

If you are teaching physics at Key Stage 3 (as well as chemistry, biology, Earth science and astronomy) and you have a physics degree, then you are a rare commodity. Department for Education and Skills (DfES) statistics show that over 70 per cent of teachers dealing with physics topics at Key Stage 3 do not have a degree in physics, and almost 40 per cent do not even have an A level in the subject. The figures for chemistry are 60 per cent and 20 per cent, whilst for biology it is 55 per cent and 35 per cent.

Teaching outside your subject specialism, even at Key Stage 3 level, can affect your confidence in tackling topics. Across a science department there is likely to be a good spread of specialisms, although an increasing number of schools are lacking any subject specialists in physics, and chemistry is following on behind. This means that it is vital that planning for progression in pupils' learning is carried out by the whole department working together.

Contemporary science

We should include more examples of contemporary science in schemes of work. This recommendation does not just come from OFSTED, but from pupils themselves. A recent study by Osborne and Collins (2000) reports that *pupils suggested that there was a need for more contemporary examples so that school science appeared, at least occasionally, to address the same issues as science in the media.* The importance of using contemporary examples has been recognised by the Key Stage 3 Strategy, which includes contemporary science as a key theme.

Contemporary science can mean several things. It can be new science – science as it is happening in the research world. It can also mean contemporary issues – news items in the media, but where the science is not necessarily new. In short, it can be 'new science' or 'science in the news'. Both are important in developing relevance in the science curriculum. New science, such as cloning, the development of new materials, and progress in space science, can be used to exemplify the way that scientists work today and how science is a living, breathing enterprise, capable of transforming people's lives. They are also useful in providing the 'wow' factor. 'Science in the news' can also be used as starting points for activities. Newspaper clippings, extracts from TV programmes, or topical web sites are all good sources of information on a wide range of contemporary issues with a science dimension.

TEACHING CONTEMPORARY SCIENCE

- Select a unit from your Key Stage 3 scheme of work.
- Identify two opportunities for the inclusion of new contemporary science contexts (new science or science in the news).
- Search out media items or other sources of information which you could use (or adapt for use) with your pupils.
- Plan a lesson based around this new context.

Scientific knowledge, as measured by the output of published research reports, is expanding at an ever-increasing rate. It is impossible for anyone to keep up with even a fraction of it. However, teachers' subject knowledge cannot stand still – there is a constant need to keep up to date, which is often difficult for hard-pressed practitioners to meet. The review SET for Success (Roberts, 2002) identified the problem:

CPD is vital for all teachers, but especially for teachers in science and technology, who must stay abreast of technical and scientific progress in order to capture pupils' interest through engaging them in contemporary scientific issues. Teachers with knowledge of the latest developments in the sciences are better able to interest science and engineering students in these subjects and enthuse them to study the subjects at higher levels.

(Roberts, 2002, p 62)

CPD (continuing professional development), including attending updating courses, taking part in bursary schemes which allow teachers to work in research settings, and working with scientists and engineers in the classroom, will help teachers to keep up to date in some ways. However, it is a career-long process, not confinable to one or two injections of new knowledge.

Controversial issues and scientific controversies

There are many issues facing society today which have a scientific dimension. Recent decades have seen a range of curriculum initiatives which have attempted to provide curriculum frameworks and guidance on 'Science and society' issues. Currently, the programme of study for science at Key Stage 3 has a number of contexts which could be used as a starting point for dealing with controversial issues.

PROGRAMME OF STUDY SECTION	RELEVANT STATEMENT FROM PROGRAMME OF STUDY
Scientific enquiry	• Interplay between empirical questions, evidence and scientific explanations using . . . contemporary examples
Life processes and living things	• The need for a balanced diet • How toxic materials can accumulate in food chains
Materials and their properties	• Recognise the importance of chemical changes in everyday situations • How acids in the environment can lead to corrosion of some metals and chemical weathering of rock
Physical processes	• Some effects of loud sounds on the ear • The distinction between renewable and non-renewable resources

This is just a selection. There are many more examples.

USING CONTEMPORARY ISSUES

- Take one of the above examples (or choose your own).
- Identify in your scheme of work where the topic is covered.
- Assess whether you deal with this topic in a way which brings out the contemporary dimension or a current controversy.
- Plan how you could include contemporary science/issues and controversy into a lesson.

Teaching about controversial issues

Controversial issues are often topical, and may connect with pupils' lives. They could relate to topics and ideas about which pupils already have beliefs and views. This means that tackling such issues can be a powerful learning experience, but also a problematic one. As an adult we may realise that there are several standpoints from which to arrive at an overall, often tentative, judgement. However, pupils may believe that there is a right or a wrong answer. This puts the responsibility on the teacher to provide a skillful combination of objectivity, neutrality and balance. Teachers are authority figures to pupils, and care must be taken to prevent the teacher's own views on an issue from being passed on to youngsters as if those views were matters of fact.

Some facts will be involved, however. In a discussion about, say, mobile phone safety, the teacher will be able to introduce science concepts where relevant. The teacher will also be able to correct mistakes in the factual basis of others' arguments. This is where being objective is important.

Some controversies involve dealing with peoples' values and beliefs, including the teacher's own. This is where neutrality and balance come into play. The teacher, when tackling an issue such as the controversy surrounding animal testing, will need to ensure they do not use the lesson merely to promote their own views. Although it is often difficult to achieve, neutrality should be maintained when dealing with such an issue. This should be matched with seeking out balance in the arguments and positions. Tensions may also arise between neutrality and balance. What if a class of pupils fails to come up with a key point in favour of a particular technological development the impact of which they are considering in a discussion? Should the teacher introduce the missing point? This would help to maintain balance, but at the possible expense of neutrality. Dealing with all this requires skill, sensitivity and awareness of the breadth of opinion in the class, including relgious beliefs, which may be important when dealing with topics such as IVF or xenotransplantation.

A recent report (Wellcome Trust, 2001) lists a series of biomedical topics which teachers currently cover in their lessons. The teachers were from a range of subjects, and were not restricted to Key Stage 3.

The table below shows that, most teachers do not deal with such issues (except for AIDs/HIV). Pupils could easily meet these topics in English, humanities or PHSE lessons, where class debates and discussions are more common than in science lessons. However, the background knowledge of the average science teacher is a vital resource when tackling such topics. Rather than leave the moral and ethical questions to others, we should seek to include appropriate opportunities for showing how science and society interact. The question is, how do we do it?

TOPIC	PERCENTAGE OF TEACHERS IN SURVEY WHO COVERED TOPIC	RELEVANCE TO KEY STAGE 3 PROGRAMME OF STUDY FOR SCIENCE
AIDS/HIV	73.3	Replication of viruses
Genetic engineering	54.9	Inherited causes of variation
Eating disorders	54.4	Need for a balanced diet
IVF	48.5	Human reproductive system
Animal experiments	44.9	Ways in which living things and the environment can be protected
Nature/nurture	43.3	Environmental and inherited causes of variation
Over-prescription of antibiotics	33.1	How the body's natural defences may be enhanced through immunisation and medicines
Genetic fingerprinting	32.1	Inherited causes of variation
Human Genome Project	25.4	Inherited causes of variation

Teaching approaches for controversial issues

It is vital that these topics are covered in a way which allows pupils to explore the various sides of the controversy. This means allowing pupils time to talk – to each other, to you, to the whole class, or to audiences beyond the classroom. Specific techniques include:

● direct teaching;
● small group discussion;
● brainstorming;
● role play;
● simulation;
● games;
● class debate;
● research and presentation.

Use of the Internet for research

A quick search of the Internet can quickly identify a large number of sources of information on most controversial issues. It is worth checking these sites before using them in your lessons to make sure:

● the content is relevant;
● the language and presentation is appropriate;
● the views expressed are not too extreme.

Many web sites present too much information. Giving the pupils a task which focuses their research helps them to select relevant information more easily. An example could be to prepare a case for one side of a debate, or to summarise views for a poster.

Some of these techniques are described in more detail in Chapter 1, Teaching, learning and assessment.

Using contemporary science and issues as starter activities

New science, or science in the news, can be used to create starter activities which will motivate pupils. They will engage with the main activity more effectively if they are 'hooked in' from the start. Contemporary starting points can demonstrate the link between what the pupil is learning, in terms of the National Curriculum content, and the world of science and technology they meet in everyday life.

Utilising pupils' natural interest, curiosity and inquisitiveness seems an obvious strategy and by using contemporary science and issues we should be able to do this. By the time new scientific developments reach text-books they are no longer cutting edge or contemporary, so it becomes vital to encourage pupils to access material from other sources. This will mean that you can increase their understanding of the importance and relevance of contemporary science by examining the applications, old and new, of established science while also discussing new advances and controversial issues.

By making available to pupils popular science journals, newspapers, etc. and encouraging them to listen to and watch news reports and TV programmes with science content as a regular part of their lessons, you will be encouraging them to take a continuing interest in the subject and showing that you recognise the importance of contemporary science. It goes without saying that the Internet is also an excellent resource for identifying contemporary science contexts for learning.

In most cases, the main activity of a lesson will focus on dealing with currently accepted scientific explanations – cell structure, energy transfer or rock formation for instance. Repeated reference to contemporary aspects of these topics may become cumbersome or artificial if maintained across an entire unit. However, the starter activity can provide a place to introduce current research, a controversial and topical theme or an example of science and technology from everyday life. Whilst these may also be relevant elsewhere in a lesson or unit, a well-constructed starter activity with a contemporary focus can provide the ideal jumping-off point for dealing with the science content to follow. The following table shows examples of starter activities based on contemporary contexts.

LESSON TOPIC	CONTEMPORARY DIMENSION USED FOR STARTER ACTIVITY
Immunisation (Sc2)	Look at the media story about MMR vaccination, and the reduction in the number of parents opting for it in the UK. Use newspaper cuttings or digital resources. Ask why the story was such headline news, leading to main activity which looks at immunity and vaccination.
Acids in the environment and corrosion (Sc3)	Brief recap of the types of fuels which have been developed over past 10–15 years (focus on introduction of unleaded petrol and then recent development of low sulphur petrol). Discussion of why this product has been developed.
Planetary motion and gravity (Sc4)	If available use a two-minute clip from the films *Armaggedon* or *Deep Impact*. Could it happen? How can we predict future impacts?

Summary

In summary, this chapter has highlighted the need to:

- plan programmes where teaching of the five key ideas are sequenced and cross-referenced so as to maximise the progressive development of pupils' scientific understanding;
- identify examples of contemporary science or issues which you could build into your teaching;
- choose teaching and learning approaches which are effective when dealing with controversial issues;
- develop starter activities that are based on contemporary science or issues.

Further reading

New Scientists

Still the best source of science news, updates and recent discoveries. Every science department should subscribe.

Driver, R et al (1994) *Making sense of secondary science: research into childrens ideas.* London. Routledge.

Based on the work of the *Childrens Learning in Science* project, this book explores the misconceptions pupils have with respect to the full breadth of topics in the national curriculum, and suggestions as to how to overcome them.

Raising standards

This chapter is about some of the things that you could do both in your classroom and across the department to help make science teaching as effective as possible. It is all about reviewing, planning and monitoring – which makes it sound a less than interesting topic, and hopefully someone else's responsibility. On the contrary, we would like to show you that:

- this area is something that all of us already build into our practice as competent teachers;
- there are some straightforward activities for you to try that could be useful in your teaching;
- by doing these things and discussing them you could be contributing to the effectiveness of the department, whilst developing your own professional skills.

The topics to be covered include:

- leadership and management for all;
- planning for departmental improvement and pupil attainment;
- finding out how well we compare with others;
- setting annual targets for pupils and monitoring their progress;
- monitoring and evaluating our teaching.

The Key Stage 3 Strategy is about improving progression, raising expectations, improving engagement and transforming teaching and learning. The topics in this chapter will have an impact on each of these areas – a clear understanding of these topics will help us see how different aspects of the Strategy fit together.

KEY ISSUES

In this chapter, you will find out how to:

- use school, local and national data and information from other sources to evaluate departmental performance and set targets at school and pupil level;
- use all available data to identify strengths and weaknesses, including those related to groups of pupils such as boys or girls, pupils with special educational needs of all kinds or of differing ethnic backgrounds, and propose strategies for improvement;
- use classroom observation and other evidence to monitor specific aspects of teaching and learning within the department and evaluate pupils' achievement and engagement.

This chapter is written on the premise that all teachers, not just curriculum leaders, will want to know about each of the activities that form the Key Stage 3 Strategy, out of interest and for their own continuing professional development.

Leadership and management for all

A competent classroom teacher exhibits considerable management and leadership expertise in their day-to-day teaching. Indeed, much of the feedback a student teacher receives from their tutor concerns the management and leadership of learning. This list illustrates the sort of activities that would fit under each of the two headings.

MANAGEMENT ACTIVITIES	LEADERSHIP ACTIVITIES
• Ordering and distributing equipment	• Having a clear sense of purpose
• Organising learning activities	• Communicating aims
• Maintaining records	• Motivating and inspiring others

If we take a step back from being a classroom teacher for a moment and think about the team of people that make up the science department, we can see that exactly the same skills are needed for the effective running of the department. The head of department has overall responsibility for leadership and management, but we would argue that large departments also require a curriculum leader with responsibility for Key Stage 3. Sometimes teachers with a curriculum responsibility for one specific science subject are also given the role of Key Stage 3 co-ordinator within the department. In our experience the time and energy needed to organise the teaching of a subject at GCSE and AS/A2 level leaves insufficient opportunity to be an effective key stage co-ordinator. Our concerns are supported by recent inspection reports which note that the day-to-day management of science departments is good, yet the quality of key stage 3 leadership varies considerably (OFSTED, 2000).

It is so important that pupils experience a coherent and effective science curriculum at the start of their secondary careers, and one which has a high status in the eyes of pupils and teachers alike. To achieve this the curriculum leader needs a clear sense of purpose and the ability to communicate this to people and to motivate them to achieve. An example of this is provided by a recent inspection report on science teaching:

> *Leadership is outstanding in enthusiasm for the subject, in ensuring clear and pragmatic principles for teaching and learning and in forging a united*

team of teachers and technicians. One of the main reasons for the success of Key Stage 3 science is the collaborative approach to the development of the curriculum. All members of the department, including technicians and support staff, contribute to the development of schemes of work and resources for learning.

(OFSTED, 2000)

The list of activities that make up departmental management and leadership is similar to the list of classroom skills at the start of this section.

MANAGEMENT ACTIVITIES	LEADERSHIP ACTIVITIES
• Reviewing and updating resources and schemes of work	• Motivating and inspiring teachers, linked to their professional development
• Organising teaching groups and staffing	• Communicating aims and progress to science teachers and senior managers
• Monitoring science teaching and assessment	• Encouraging debate on effective teaching and learning
• Monitoring the progress of the department against set targets	• Having a clear sense of purpose • Building effective relationships with others (curriculum areas, support staff, primary schools)

We all possess these skills to varying degrees and should be encouraged to use them to contribute to the effective organisation of science teaching within the school. Leaders will often delegate tasks to others. This helps to spread the workload, but also helps motivate individuals, increases job satisfaction and builds stronger teams. If your curriculum leader is not strong on delegation, then volunteer!

Leadership and management in the Key Stage 3 Strategy

An initial activity in the Key Stage 3 Strategy is the departmental audit. Curriculum leaders will be expected to review the teaching and standards of the department, identify targets and work with staff to identify their professional development needs. They then need to support the implementation of the strategy by leading the training and monitoring its implementation.

Specific management and leadership activities are the focus of part of the Key Stage 3 Strategy audit. Some of the criteria are reproduced below to encourage you to have a go at this activity in your own school. You may also like to identify your own strengths and priorities for development.

CRITERION	FULLY IN PLACE	PARTLY IN PLACE	NOT IN PLACE
Targets set for pupil attainment at Key Stage 3.			
Plan in place for how to achieve these targets.			
Plan monitored by the curriculum leader and senior management of the school.			
Teachers receive feedback when senior managers monitor the work of the department.			
Monitoring linked to a programme of professional development.			
Lesson observation with feedback used to monitor the quality of teaching.			
Regular support in place for planning processes.			
Standards of pupils' work are moderated.			
Standards of marking are monitored.			
Regular departmental discussions take place about the quality of teaching and learning.			
Planned programme of professional development in place.			

Departmental strengths	
Priorities for development	

(taken from *Auditing a subject in Key Stage 3*, DfES, 2002 (b))

REVIEWING YOUR MANAGEMENT SKILLS

- Look at a copy of your job description or a generic school one for that of a classroom teacher.

- Make a list of two columns. In the first column write what you believe to be the key areas that make a good classroom manager (use examples from earlier in this chapter to get you started).

- In the second column rate yourself on a scale of 1–4 for each of your identified areas. 1 equates to novice level through to 4 as expert.

- From this you can identify both your strengths and areas that you may like to develop further in the future.

Planning for departmental improvement and pupil attainment

Many teachers enjoy the planning sequences of lessons that address specific topics in the science curriculum. We can even get quite excited when a burst of inspiration results in a different way of approaching a difficult topic. Usually the time invested pays off in terms of pupil learning, and personal and professional satisfaction with the lesson outcomes.

Fewer teachers enjoy what we would call 'administrative planning', which includes the writing of development plans. One of our authors' early experiences in this area were in response to changes in the funding arrangements in their school. 'It had been decided that the department would be allocated its capitation on the basis of a development plan for science. My first attempt did not result in the increase in budget I had asked for, which left me frustrated and thinking no one had read my very detailed and well-argued document.'

These experiences, and a lot more practice, can teach us that there is more to planning than simply producing a document.

- There are different levels of plans, each with different purposes, target audiences and amounts of detail. Identify which sort of planning you are doing.
- Creating plans helps clarify ideas and identify priorities for groups as well as individuals. It can be an effective team building strategy. Do it with other people, not alone.
- Good planning, like good teaching, involves monitoring, reviewing and developing. Build these activities into your plan.

Here are some activities associated with three different levels of planning.

	ASSOCIATED WORDS	PURPOSE	AUDIENCE	LEVEL OF DETAIL
Short-term	Lesson plan	Working document	Restricted to you and one or two others	High
Medium-term	Scheme of work, action plan	Provide a framework	Teams of teachers	Medium
Long-term	Development plan	Identify priorities	Team and beyond	Low

---

CASE STUDY — THE THREE LEVELS OF PLANNING WHEN ORGANISING A FIELD TRIP

Produce a written proposal to go to the head and governing body, seeking permission for the trip. Include purpose, dates, names of those involved but few details at this stage.

Long-term plan – brief, shared agreement, 'outside audience', could be altered, but this would be difficult once agreed. The arrangements include booking transport, informing parents, arranging insurance, organising cover for teachers, informing other staff and planning the content. A considerable amount of documentation is involved at this stage.

Medium-term plan – provides a structure for the event, includes more detail and involves a number of different audiences. Documentation needs to be carefully preserved. Final details ensure everyone arrives at the agreed time with the necessary equipment and a clear idea of what is expected.

Short-term plan – the audience is now restricted to those taking part. Documentation may be necessary, but has less status than with a medium-term plan. This level of detail is flexible enough to be changed in the light of changing circumstances.

Principles of planning

Planning lessons always seems more straightforward to me than writing action plans, so let's just reconsider what goes into a lesson plan. From this it should be possible to extend similar principles into administrative planning tasks.

A lesson plan should:

- be clear about what the pupils are expected to learn;
- state the teaching and learning activities;
- list the resources needed;
- indicate how to assess the learning that occurred;
- suggest what you might change in future lessons (to be completed after the lesson).

The principles of planning departmental developments are the same – it is just a different set of words that are used.

Many teachers write lesson plans as a list of headings on a side of A4, which can easily fit into a file, so that they are easy to access, read and refer to during lessons. Such a plan is a working document that will probably get updated with margin notes as it is used. We would recommend the same format for action plans, for similar reasons.

LESSON PLANS	ADMINISTRATIVE PLANS
Lesson objective, learning objective (QCA), lesson aims	Objective, SMART target, goals
Lesson activities, teaching activities (QCA)	Strategy, actions
Resources	Costs, time, personnel, deadline, responsibility
Learning outcomes (QCA)	Success criteria, performance indicator
Improvements, evaluation	Effectiveness, evaluation

You are likely to encounter planning activities in several areas of the Key Stage 3 Strategy. Some of these are described in the next section. Before moving on there are two further points to make:

1. If you find it difficult to clearly state activities or assessable outcomes, try to break down your objectives into smaller or more manageable pieces. In planning jargon, make sure your objectives are SMART (specific, measurable, attainable, results-focused, time-based).

2. Make sure your plan communicates priorities, purposes and actions. It needs to be discussed and amended, not filed away and forgotten.

Planning activities in the Key Stage 3 Strategy

The Key Stage 3 Strategy kicks off with an audit of the department to establish a set of priorities for development. The curriculum leader is then responsible for developing an action plan that amongst other things will include training activities for individual teachers. We suggest that you will want to be closely involved in this planning activity as it offers great potential for personal development. Review periods are built into the Strategy to make sure the plan is monitored and evaluated.

Improving progression in learning, and transforming teaching and learning are key aims of the Key Stage 3 Strategy. Monitoring schemes of work (medium-term plans) and reviewing the structure of lessons (short-term plans) are essential parts of this, and are covered in more detail in the section on teaching and learning.

Schools are required to set targets for Key Stage 3 attainment in mathematics, English and science. We strongly recommend that science teachers adopt a planning strategy for this purpose that has been agreed by the whole school. A good place to start would be to use or adapt the DfES five stage development cycle, which is described at the end of this chapter.

MANAGING CHANGE

Good management of change needs good preparation to help things run smoothly.

When you are about to embark on something new it is worth making a simple analysis of what potential benefits and pitfalls there may be in implementing that change.

- Make a two-column table headed 'pros' and 'cons'.

- Using your own judgement and prior experience identify what you think will be the potential benefits ('pros') and what may be the stumbling blocks ('cons') for the change you wish to make.

By doing this you should be able to help anticipate implications of making the change.

Good planning avoids the need for fire-fighting tactics. This approach can be applied to all types of changes from small ones such as trying out a new teaching style or techniques, to larger whole school issues such as whole school curriculum timetable changes.

Finding out how well we compare with others

Pupils are spending increasing amounts of time completing assessment tasks and teachers are spending more time preparing pupils for these tasks, or marking them, than ever before – but are we getting good value from these activities? As this quotation shows, we could make more use of the results:

> Procedures for the assessment of pupils' attainment in science at Key Stage 3 are good in nearly two thirds of all schools but the use of assessment information to inform curriculum review is good in only one third. As much as ten–fifteen per cent of teaching time in Key Stage 3 science is devoted to assessment, though in general only a narrow range of techniques is used and the use made of the data is often limited. For example, in only a small proportion of schools is formal assessment used to track individual pupils' progress or judge the effectiveness of teaching approaches. The use of so much teaching time on testing is therefore often unproductive.
>
> (OFSTED, 2000)

In this section we will show how, for a small investment of your time, you can get some useful information from the wealth of assessment data that is generated in schools.

Using data to look at the past performance of the department

GETTING STARTED

The Autumn Package is sent to schools in October. It contains national performance data, ideas for how to use the data and some activities to help teachers extract information from their figures. Packages are specific to key stages, and contain data from one year's national assessments. The Autumn Packages for other key stages and years are available on the Standards web site as pdf files, for which you need *Adobe Acrobat Reader* software to read. A CD-ROM version is also available.

The Autumn Package gives you access to information on:

- how well the school performed this year;
- how well your school did compared with the rest of the country, or with similar schools;
- how well the department did compared to the rest of the school, or previous years;
- how well subsets of your pupils did compared to the rest of the year group;
- how well your class did, individually and collectively (value added);
- what younger pupils are likely to achieve in the future (target setting).

Some answers to the questions at the top of this list are provided in the school's PANDA (Performance and Assessment) report, which arrives several weeks after the Autumn Package. The rest of this section shows how you can find answers to the questions that are relevant to the pupils in your Key Stage 3 teaching groups.

The contents of the PANDA report are being used increasingly in departmental reviews, and may be used to provide evidence in the performance management review cycle. For this reason it is important that all teachers should have ready access; copies are available online.

SOURCES OF DATA FOR YOUR SCHOOL

- The Autumn Package is available from the Standards web site in October each year. A copy is also posted to each school.
- The PANDA report, which is specific to your school, is posted to schools in November.
- Results from other tests, such as NFER CAT scores, may be available.
- Your LEA will have generated data.
- Key Stage 2 data from primary schools should be available – raw test scores are particularly useful.

SOURCES OF INFORMATION AND ADVICE

- *Using performance data* (DfES, OFSTED, QCA) **www.updata.org.uk**
- Association of Assessment Inspectors and Advisers **www.aaia.org.uk**
- *Magic Markbook* **www.tes.co.uk/online/assessit**
- *Recognising progress – getting the most from your data* **www.standards.dfes.gov.uk/performance/ap/index.html**

HOW TO USE THE AUTUMN PACKAGE TO REVIEW PAST PUPIL PERFORMANCE

Value-added charts are a straightforward way of finding out if pupils did better or worse than the average. The charts take account of prior attainment, and can be used for individual pupils, groups of pupils (say from one particular feeder school, ethnic group or gender) or classes.

Value-added charts are produced by comparing two sets of assessment results for a single cohort of pupils. The charts in a Key Stage 3 Autumn Package compare Key Stage 2 points score with Key Stage 3 points score. It is possible to construct your own value-added charts with other sets of data, although large sample sizes are needed.

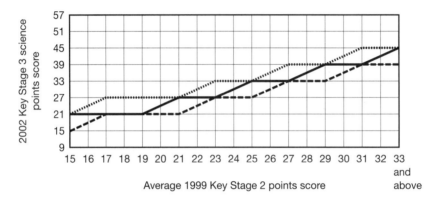

Key Stage 3 Science test level value-added line

The solid line shows the average performance of pupils nationally in Key Stage 3 assessments for 2001. Dotted lines show the limits within which a pupil's performance should be considered normal. The line has 'steps' because levels were used; raw test scores would produce a smoother line.

Example

In 1998 a pupil achieved a level 2 for Key Stage 2 mathematics, level 3 for science and level 4 for English. That would give an average point score of 21. Three years later the same pupil achieved level 5 for science (33 points). From the chart you can see that this pupil performed above normal limits compared to other pupils nationally. Was this due to exceptionally good teaching or was the pupil very hard working? This chart will not answer those questions. However, if the same pupil had achieved a level 3 for Key Stage 2 mathematics and not level 2 (average point score moves from 21 to 23), their science performance for Key Stage 3 would only be considered normal.

This example points out some important principles.

- The accuracy of the information you get out of this exercise will depend on the accuracy of the data that is put in; no tests are absolutely reliable.
- Be more concerned with general trends and patterns, perhaps over a number of years, rather than individual cases.

To use the chart you will need to convert three Key Stage 2 results and Key Stage 3 science levels for each pupil into points (use the Ready Reckoner in the Autumn Package or the table in the next section for this). By plotting points on the chart it is possible to see how well your pupils did compared to other children in the same tests. When individual pupil attainment is plotted on the chart those that fall below the bottom line have made less progress than expected; those at or above the top dotted line, better progress. The further from the average line, the more exceptional the performance. Different groups of pupils could be identified, perhaps by using different colours, to identify more precisely whether there are implications for the teaching of different groups. Using a spreadsheet simplifies this task and allows direct comparisons to be made with previous results.

GETTING A PICTURE OF WHERE YOU ARE

Using this chart it is possible to:

- look at the general performance of the department against national averages;
- compare pupil progress for different ability levels, for example comparing the value-added for level 6 pupils against value-added for level 3 pupils – by plotting the two groups on the chart but in different colours
- compare the performance of girls against boys
- find out if different science sets make similar progress.

The best time to review results is immediately after the arrival of the Autumn Package. Class teachers should analyse their own pupils' results, as they will have insights into pupil performance that may need to be considered. The reviews for each class should then be pooled to give an overview for the year group and the outcomes used to evaluate development plans. Do not let this process get too serious! Maintain a sense of perspective.

Using data to set targets for pupils and monitor their progress

Did you read the section above concerning different levels of planning? If not, this next bit may not make too much sense. You may feel that you are caught in the middle of a conflict between 'bottom up' target setting, such as helping a pupil to take the next small step in understanding a difficult idea, and 'top down' target setting expressed in percentages of pupils achieving level 5 and above. I think it is helpful to see target setting in terms of three different levels as we did with planning:

Long-term targets are broad descriptions of percentages of pupils achieving a specific level. They concern performance levels nationally, or for local authorities. 'National' cut off points are currently level 5+ at Key Stage 3 and five or more grades C and above at GCSE.

Medium-term targets are for individuals, classes or departments. They often contain a more detailed analysis and are usually reviewed and updated annually. These targets are often published for various audiences, and would typically be reported to parents. This next section relates to medium-term targets.

Short-term targets contain very specific detail but are not widely publicised. Their life cycle may be a matter of days or weeks. It is unlikely that these targets are expressed in terms of levels or grades. This aspect of target setting gets a thorough airing in other sections of the book.

Why set targets?

Setting medium-term targets is a realistic means of raising expectations of pupil attainment. It also provides useful information to track progress and improve teaching. I found it takes about an hour of classroom time to establish targets for a normal sized teaching group. The second half of the first term in Year 7 seems to be about the best part of the year – the pupils are settling into routines and have developed relationships with their teachers (and subjects!). Monitoring progress against the targets then usually takes place in the summer term, but this may vary according to the school's assessment and reporting cycle.

What do you need?

Progress charts are found in the National Value Added Information section of the Autumn Package, and provide one means of helping teachers and pupils to set individual targets at Key Stage 3. They are based on national performance data for pupils who completed their Key Stage 3 assessments in the previous year, and help to set targets in a context of national performance based on prior attainment.

Prior attainment at Key Stage 2 across English, mathematics and science (average point score) is the best predictor of attainment in science at Key Stage 3, just as the average points score at Key Stage 3 is the best predictor for GCSE science. Other test scores, such as CAT scores, provide useful additional data to consider alongside the Autumn Package. Different forms of data may produce conflicting information about an individual pupil, which can raise interesting questions about their progress.

ANYTHING ELSE?

To use the charts with your class, you will need their Key Stage 2 results for assessments in English , mathematics and science, and copies of the progress charts for science that can be projected onto a screen. A digital projector does this very effectively if you have downloaded the Autumn Package onto a computer.

Procedure

1. Convert Key Stage 2 levels to a points score (this is a lot easier if you use a spreadsheet).

SUBJECT	KEY STAGE 2 LEVEL	POINT SCORE	AVERAGE POINT SCORE
English			
Mathematics			
Science			
Total			

LEVEL		POINT SCORE
Absent	A	Disregard
Disapplied	D	Disregard
Working below	B	15
Below threshold	N	15
Level 2		15
Level 3		21
Level 4		27
Level 5		33
Level 6		39
Level 7		45
Level 8		51
Exceptional Performance (EP)		57

2. Calculate the average points score for the three subjects and then make sure each individual knows their own average point score.
3. Identify the appropriate chart for use with each point score range. Pupils

will need help with this, and so may some staff! The symbols used in each chart heading are not straightforward. See the example in the case study that follows.

4. Explain that each chart shows the levels achieved by last year's pupils in their Key Stage 3 assessments. The majority achieved the level shown by the tallest bar, but there was a range of levels achieved.

5. Ask pupils to identify the level they aim to achieve in science based on the evidence in the chart. The tallest bar is a fairly safe target, but pupils should be encouraged to consider levels above the 'mode'.

6. Record each pupil's target level. You may wish to challenge some pupils about their decision. It is a delicate balance between a target that is challenging, and one that is comfortable. We do not believe all pupils should have challenging targets in every subject. Learning should be fun too! These records will need to be passed on to other Key Stage 3 teachers in future years. A centralised recording system is essential.

7. Aggregating the results of this exercise provides the department with data on prior attainment for the year group, and also a measure of expected performance at Key Stage 3.

Recording

The pupils also need to record their targets; planners or diaries are an effective means of providing parents with this information. A computerised system is essential for tracking pupil progress in a large school. It helps to create a common approach to recording and reporting, usually improves accuracy, and makes information retrieval much simpler. *Assessment Manager* from SIMS is very effective, but requires a whole school commitment. Otherwise a spreadsheet would be adequate for departmental use.

Monitoring

Predicting an outcome for the end of Key Stage 3 based on Key Stage 2 performance is an important step in raising expectations. Progress towards targets needs to be monitored – giving feedback to the pupil provides a strong motivational force; feedback to the teacher enables teaching plans to be adjusted. Again, we recommend a whole school approach to monitoring, with specific points identified on the school calendar to coincide with reporting procedures.

There is an assumption that pupils will make steady progress over the three years, and that each level represents two years' progress. If you want to set interim targets for each year it may be helpful to sub-divide each level. Sub-levels are used extensively in primary schools to help set targets and track progress by teacher assessment. The sub-levels used are a, b and c, which represent high, medium and low attainment within the levels. Thus a pupil who is nearly achieving level 6 would be considered 5a. Each pupil would therefore be expected to progress by one sub-division every two terms. Using mark thresholds for the Key Stage 2 science assessment would be one way of establishing sub-levels at the start of Year 7.

The *Test Base* CD-ROM from the DfES is a useful source of assessment material if teachers wish to use tests as a means of monitoring progress towards target levels. Optional tests for Years 7 and 8 are available from the QCA.

CASE STUDY — YEAR 7 INTAKE DATA

NAME	TEST LEVEL AT KS 2			AVERAGE POINT SCORE	KS 2 SCIENCE TEST MARK	SCIENCE SUB-LEVEL	PREDICTED KS 3 LEVEL
	En	Ma	Sc				
Brian Able	5	5	4	31	44	4c	6b
Lucy Lowe	3	4	4	25	63	4a	6b

The average point score was calculated using the Ready Reckoner, although it is easier to set up a spreadsheet to do this task. Using the mark range thresholds it is possible to use raw Key Stage 2 results to arrive at science sub-levels. The Key Stage 3 predicted level was calculated from the appropriate progress chart.

The progress chart for pupils with a Key Stage 2 average point score of between 23 and 25 was used to calculate Lucy's predicted Key Stage 3 level. A target of 6b is a considerable challenge for Lucy as only 12 per cent of pupils with her prior attainment achieve this level nationally. However, she performed very well in her science test at Key Stage 2 and enjoys science, so Lucy felt that this was a realistic target. A different progress chart (average point score of 31) shows that a target of 6b for Brian would be less challenging as the majority of pupils with his prior attainment achieve level 6 at Key Stage 3.

Monitoring sheet

NAME	KS 2	YEAR 7			YEAR 8			YEAR 9		
		Predict	Actual	Diff	Predict	Actual	Diff	Predict	Actual	Diff
Brian Able	4c	4a	4b	−1	5b			6b		
Lucy Lowe	4a	5c	5b	+1	5a			6b		

The monitoring sheet shows that Brian will have to make more progress throughout the key stage if he is to achieve his target. This was discussed with him at the time the target was agreed at the start of year 7.

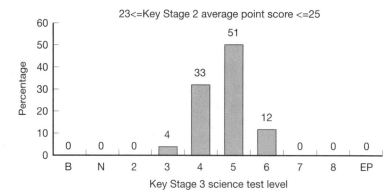

Points score graphic

Monitoring and evaluating your teaching

The Key Stage 3 Strategy is about improving progression, raising expectations, improving engagement and transforming teaching and learning. Any improvements are going to be difficult to detect unless we look in detail at things happening inside classrooms. From the point of view of a pupil, we would want to know:

- What goes on in science lessons?
- What other science activities can I do outside lessons?
- What does the teacher think of me as an individual?
- What do I think of science?
- How well am I doing?

You may argue with some of the questions in this list, but the principle is clear – any effective evaluation will need to look at a wide range of evidence, and will focus on what is taught, the quality of the teaching and the effect it has on the pupil. Lesson observation is an essential approach.

Monitoring and evaluating

Monitoring is about the collection of evidence. Evaluation is a consideration of the evidence to assess effectiveness. Teachers in the department developed targets and plans as a collaborative effort; they should all be involved in the evaluation process too. But who should do the monitoring?

- Monitoring helps maintain the momentum of the initiative by focusing on priorities.
- Monitoring helps the monitor reflect on their own practice.
- Monitoring helps spread good practice.

We would encourage all colleagues to share in a range of different monitoring activities.

It is worth noting that monitoring can make people nervous, particularly lesson observation. Always try to give the person observed some feedback immediately after the lesson, and create an opportunity for longer discussion soon afterwards. Giving and receiving feedback is quite a skill – it needs to be positive and supportive, but must not avoid identifying areas for development where appropriate.

SOURCES OF EVIDENCE

- Lesson observation.
- Sampling pupils' work.
- Assessment material – formal, such as tests, teacher assessments; and informal, such as pupils' self assessment. For pupils who could attain above level 7 in terms of Teacher Assessment, QCA have produced a set of Extension Tasks, available from the QCA website.
- Documentary evidence including planning sheets, work sheets, equipment requisitions.
- Use of rewards and sanctions.
- Discussions with pupils, teachers and parents.
- Pupil feedback through surveys or questionnaires.
- Other professional sources, such as performance management evidence, NQT observation and OFSTED material.

There is no right answer as to which sources of evidence to use. The evidence that you need will depend on the targets that you set. You also need to discuss the criteria that will be used and the methods used for recording.

POSSIBLE CRITERIA

These criteria are based on the OFSTED framework:

- quality of teaching of science content;
- organisation of the lesson;
- responding to pupils;
- management of learning and relationships.

You could break down each heading into more specific aspects – as shown in the next table.

Other sources of possible criteria include the Key Stage 3 audit document (*Auditing a subject in Key Stage 3, DfES, 2002 (b)*)

Monitoring planner

The following table can be used as part of your planning when deciding which sources of evidence will be used for monitoring.

Examples of criteria	TEACHING Engages pupils	TEACHING Questioning technique	ORGANISATION Clear lesson Objectives clear	ORGANISATION Assesses progress	INTERACTIONS Response to pupils	INTERACTIONS Set targets for improvement	MANAGEMENT Pupil involvement	MANAGEMENT Feedback on progress
Lesson observation								
Work sample								
Assessment material								
Documents								
Rewards and sanctions								
Discussion								
Survey								
Other professionals								

Identifying which sources of evidence you will use for each criterion.

Other monitoring activities

- Reviewing the scheme of work, including the role of ICT.
- Reviewing teaching materials such as textbooks, worksheets, videos and software for coverage of the scheme of work, readability and quality.
- Reviewing support systems for newly qualified and non-specialist staff.
- Reviewing the deployment of teaching staff to Key Stage 3 classes.
- Reviewing the deployment of support staff.

The next section looks at how all these areas can fit together.

ASKING YOUR PUPILS

Making the most of your pupils' opinions, take a snapshot of their perceptions of a topic. Try this with a group that you feel established with.

- At the end of a completed topic ask your pupils to rate both their level of understanding of the topic and the level of effort that they have made.

- This is best done by making bar charts, the Y-axis marked from 0–10. The pupils enter their own perceived scores as columns for both level of understanding and level of effort in the topic. This can be a useful diagnostic tool for both teachers and pupils. Patterns and connections between pupils' perceived level of understanding and effort can sometimes be revealing.

- With an established group this activity can be worth repeating over a number of topics, accumulating data on the same graph over a period of time.

- To take this activity one stage further try asking pupils to write a brief explanation of why they marked their perceived understanding and effort in the way that they did.

Summary

This chapter examined some of the things that you could do to help make science teaching as effective as possible. We have looked at leadership, planning, comparing results, setting targets and monitoring. Any one of these aspects should contribute to an improvement in teaching and learning, but to be really effective they need to be fitted together into a coherent package. The activities may be linked together into an annual cycle, and if planned carefully can be used to make a significant contribution to the school's programme of assessment and reporting.

Integrating these activities into a coherent programme of departmental improvement

The DfES has produced a suggested five-stage cycle for whole school improvement. We would certainly recommend trying this as a model for the science department to use as you work through the different phases of the Key Stage 3 Strategy. The timing of some stages is controlled by outside events, such as publication dates of results and the distribution of the Autumn Package. There are also statutory requirements for the school to publish targets for Key Stage 3 attainment in the autumn term. Perhaps the stages of the cycle could be published on the school calendar. The timings suggested in the table that follows assume the cycle starts at the beginning of the school year – but do not let this prevent you from starting at any other time!

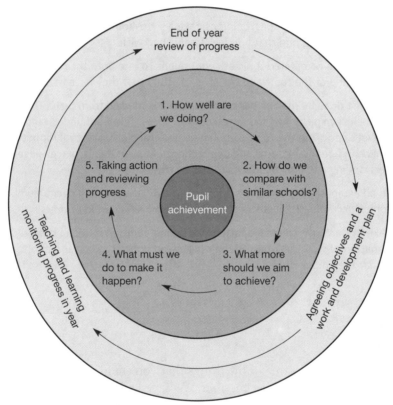

The five-stage cycle for school improvement

The table shows how the activities in this chapter relate to the objectives of the Key Stage 3 Strategy and how they could fit into an annual cycle. The model chosen here was developed by the DfES to help schools organise the use of assessment data to raise pupil achievement. You may wish to adapt it or develop your own framework.

Summary table to show the links between DfES five-stage cycle, Key Stage 3 Strategy objectives and the sections of this chapter

FIVE-STAGE CYCLE	ACTION	TIMING	RESOURCES	SECTION IN THIS CHAPTER	KEY STAGE 3 STRATEGY OBJECTIVES
How well are we doing?	Analysing the current achievement of pupils	End of summer term, start of autumn term	Aggregated and individual results		Use all available data to identify strengths and weaknesses, including those related to groups of pupils such as boys and girls, pupils with special educational needs of all kinds or of differing ethnic backgrounds and propose strategies for improvement.
How do we compare with similar schools?	Comparing results with previous cohorts, other schools in the LEA and nationally	October	Autumn Package Key Stage 2 & 3 results for current year 10	'Finding out how well we compare with others'	
What more should we aim to achieve this year?	Setting realistic targets for individuals and the department	November	Autumn Package Key Stage 2 results, test scores for current Years 7, 8, 9	'Setting annual targets for pupils and monitoring their progress'	Evaluate departmental performance and set targets at school and pupil level.
What must we do to make it happen?	Revision of development plans at school level, creation of medium-term plan for department after identifying appropriate actions	By December	Previous plans, targets	'Planning for departmental improvement and pupil attainment'	Evaluate the effectiveness of staff deployment in the department. Have strategies for analysing and evaluating learning resources in relation to subject and pedagogical knowledge to determine which best support the intended learning and plans for reviewing, developing and organising departmental resources.
Taking action and reviewing progress	Implement plans and monitor progress	Spring and summer	Monitoring record sheets	'Evaluating and monitoring our teaching'	Use classroom observation and other feedback to monitor specific aspects of teaching and learning within the department and evaluate pupils' achievement and engagement.

In summary, this chapter has highlighted the need to:

- contribute to planning activities in the Key Stage 3 Strategy;
- explain the purpose of the Autumn Package and the PANDA report;
- use data to set targets for classes and individuals;
- create a tracking system for pupil progress;
- contribute to the monitoring and evaluation of teaching in the department.

Further reading

QCA leaflet (2000) *Teachers use assessment for learning to raise standards* (a very useful and brief summary giving implications for classroom practice).

Bet McCallum (2000) *Formative assessment: implications for classroom practice*, Institute of Education, London (very readable booklet which reviews recent research into assessment and how the conclusions can be translated into improved classroom practice).

Neesom Report for QCA (2000) *Report for teachers' perception of formative assessment: Teachers use assessment for learning to raise standards* (the QCA leaflet is discussed in this report).

The above are available on the QCA website:
www.qca.org.uk/ca/5-14/afl/resources.asp

Shirley Clarke (2001) *Unlocking Formative Assessment*, Hodder and Stoughton (although based on work in primary schools this book provides insights and strategies which are equally applicable to secondary teaching).

DfES leaflet *From targets to action* Guidance for schools on how to set effective targets, utilising the five stage school improvement cycle.

DfEE leaflet (1997) *Setting targets to raise standards* This leaflet describes the target setting process at school level and provides a broad range of case studies of target setting in action.

The above two leaflets are available on the Standards website:
http://www.standards.dfes.gov.uk/otherresources/publications/?verson=1

Section 2: Introduction

This section of the book offers practical support to teachers by suggesting activities to provide more opportunities for effective teaching. It builds on six units from the QCA Scheme of Work for Key Stage 3 and, in many cases, provides alternative strategies to those provided by the QCA alongside a 'points to note' section based on issues raised by the authors.

Each unit is set out as a grid under the following headings:

- learning objectives;
- possible teaching activities;
- learning outcomes;
- points to note.

The units are cross-referenced to some of the key issues raised in Section 1 and each unit includes a more in-depth treatment of one or more of these issues. The learning objectives and learning outcomes are either taken directly from the QCA Scheme of Work or are summarised.

The units

The units were selected to provide:

- two units from each of Sc2, Sc3 and Sc4;
- examples from early, mid and late Key Stage 3.

In most cases, the matching of a unit with a particular issue was determined by the relevance of the issue for that unit. For instance, areas of Key Stage 3 science which contain significant examples of common pupil misconceptions include electricity and plant nutrition, and so units 7J (Electrical circuits) and 9c (Plants and photosynthesis) were chosen to exemplify this issue.

The selected units and key issues are as follows:

TITLE	QCA SCHEME OF WORK REFERENCE	ISSUES
Cells	7A	Differentiation, contemporary science
Electrical circuits	7J	Numeracy, misconceptions
The solar system and beyond	7L	Learning resources, progression (Key Stage 2 to Key Stage 3)
Atoms and elements	8E	Assessment for learning
Plants and photosynthesis	9C	Misconceptions
Environmental chemistry	9G	Literacy, teaching and learning, contemporary science

You may find it helpful to read through the six units when reviewing your own scheme of work. As many schools' schemes are based on the QCA approach, cross-referencing to the units here should be relatively simple.

Using the units as the basis of a review should also include consideration of the issues we have raised. These sections are meant to provide 'food for thought' as you select approaches and strategies which address such matters as the inclusion of contemporary science examples, approaches to differentiation, and building in opportunities to develop literacy and numeracy skills through the science curriculum. What we have written is intended to stimulate discussion about these issues in the context of the Key Stage 3 scheme of work as a whole. Principals described here can, and should, be applied to the whole of the Key Stage 3 curriculum for science.

Cells

(Year 7, from QCA unit 7A)

Ideas for activities

This unit presents one of the key ideas essential for an understanding of many other biological phenomena. It relates to a number of topics which would have been covered at Key Stage 2, including work on micro-organisms and reproduction, where the idea of cells may have been introduced. Many topics met later in Key Stage 3 require an understanding of the cell concept, such as nutrition, reproduction, respiration and genetics. The unit provides opportunities to engage pupils in a diverse range of practical techniques, particularly microscopy.

LEARNING OBJECTIVES	TEACHING ACTIVITIES	LEARNING OUTCOMES	POINTS TO NOTE
Understand that plants and animals contain organs, and that these are made of tissues	Identify examples from everyday life and language where pupils may have met references to human organs by name (e.g. brain teaser, heart breaking, liver transplant) Discuss function of these organs Use of video (e.g. TV programme about health issues or transplant surgery; foetal images from ultrasound) to introduce idea of human body being made up of organs Display of edible 'fruit and vegetables' to include examples of roots (e.g. carrots), leaves (e.g. cabbage) buds (e.g. sprouts), fruit (e.g. tomato, apple), seeds (e.g. nuts, beans) and stems (e.g. potato, celery) to introduce idea of flowering plants being made of organs. Compare with whole plant, e.g. geranium Use slides to show structure of some organs, showing that organs are made up of different tissues	Identify, locate and describe function of a range of human organs Make suggestions about the structure of living things and about the functions of organs Living things are made up of tissues and organs, and that tissues are made of very small units called cells	Literacy: opportunity to connect scientific vocabulary with everyday use of words. Continuity: pupils will have met some organs and their functions at KS2, and will have met the life processes common to living things. Teaching and learning: have as varied a selection of media and resources showing organs, tissues and cells as possible. Misconceptions: pupils sometimes mix up names and concepts relating to 'very small things' and may confuse cells and molecules. Direct observation of real cells will help to develop the cell concept.

LEARNING OBJECTIVES	TEACHING ACTIVITIES	LEARNING OUTCOMES	POINTS TO NOTE
	Show substructure of a tissue and introduce the idea of a cell		
Use a microscope safely and effectively Produce slides containing specimens for microscopic examination Make observations using a microscope and record the observations as drawings	Have a range of microscopes and hand lenses set up round room with common objects to view (e.g. hair, sand, newsprint, flour, salt) indicating magnification of each Allow pupils to practice altering focus. Have several with eye-piece graticules Pupils make slides to show sand, flour or alternatives and draw their observations.	Describe how objects appear under magnification Correctly focus microscope Prepare slides to view under microscope Make drawings of objects being observed	Getting pupils to identify structures using a microscope is difficult. If possible, use a projecting microscope, or have microscopes set up with drawings next them showing what pupils should be focusing on when they observe a slide. **Numeracy**: pupils could use eye-piece graticules to estimate the size of objects observed under the microscope.
Understand that ideas about the structure of living things have changed over time Know that plants and animals are made up of cells	Show pupils examples of early magnified images, e.g. Hooke, Pasteur Explain how knowledge of structure of living things has developed as magnifying technology has improved Use models and photomicrographs to introduce the term 'cell' Pupils make slides of plant material (e.g. onion skin) Pupils make slides of their own cheek cells, using approved procedure Demonstrate with OHT, micro-projector or computer how to identify cells	Describe some earlier ideas about the structure of living things Explain how evidence from microscopic observations has led to new ideas about cells, and how developments in magnifying technology have assisted in refining ideas about cells Know that living things are made of cells Draw cells from direct observation	**Science investigations**: pupils could attempt to make their own mini-microscope using loop of wire holding a droplet of water. **Ideas and evidence**: the link between existing knowledge, observation and the development of new ideas about cells.
Understand that plant and animal cells show some similarities and some differences Understand that animal and plant cells have a surface membrane, cytoplasm and a nucleus, and the functions performed by these structures	Compare structure of animal cell and plant cell using direct observation and demonstrations (e.g. photomicrographs) Construct chart to show the similarities and differences in animal and plant cells Make models of plant and animal cells using	Identify observable differences between animal and plant cells Relate parts of model cells to diagrams and pictures of plant and animal cells	**Progression**: many topics covered in the Key Stage 3 Scheme of Work require a good understanding of cells.

LEARNING OBJECTIVES	TEACHING ACTIVITIES	LEARNING OUTCOMES	POINTS TO NOTE
Understand that plant cells have a cell wall and a cell vacuole and that some plant cells (e.g. in leaves) contain chloroplasts	plastic bags, cellulose paste, modelling clay, boxes, etc. Explain what the cell membrane, cytoplasm and nucleus do in all cells, and what the cell wall, vacuole and chloroplasts do in plant cells		
Understand that there are different types of cells, each adapted for a different function	Use prepared slides or photographs from the Internet showing range of cell types to introduce pupils to variety of cell structures Show pupils examples of different types of cell (e.g. photomicrographs, drawings, mounted specimens) and get them to identify what their function is from range of possibilities	Understand that different types of cells can be found in plants and animals and that these cells carry out specialised functions Relate structure to function in selected cell types	**Learning resources**: the Internet contains many sites which provide images of cells and tissues. **Teaching and learning**: use a card game with pictures of cells to be matched with descriptions of the cell's function.
Understand that cells form tissues and that tissues form organs Name some important tissues in plants and humans Explain the organisation of tissues	Use modelling clay to make a tissue from building blocks of cells Build layered structure of an organ such as the stomach Use photomicrographs, etc. to identify a range of tissues and their functions	Identify and name some examples of tissue types from plants and animals	

Key issues

Differentiation

As this is a Year 7 topic it is likely that you will be teaching it to a class that has not been segregated according to attainment. It is important therefore to develop teaching approaches that keep all pupils involved and suitably challenged by a range of tasks. Regardless of the ability spread in a class, there will be a host of other ways in which the class will vary, from pupil to pupil. Examples include literacy and numeracy skills, pupil interest, gender, ethnicity and degree of parental support.

When dealing with the whole class, aim to get maximum involvement through graded questioning and begin by asking questions that are accessible to all pupils. In setting tasks, bear in mind the 'expectations' section of the

QCA Scheme of Work. This describes what to expect from most pupils, from those who do not progress as quickly, and from those who progress at a faster rate. Skillful use of these statements will allow you to set up activities which differentiate by both outcome and task.

DIFFERENTIATION BY OUTCOME

This is where a common task is set, but where some pupils will demonstrate a greater understanding than others in their learning outcomes. For instance, in a lesson where pupils are observing and drawing cells, pupils who have progressed further than most could produce labelled diagrams of onion epidermis cells where cell parts are clearly labelled and cell size is estimated. Whilst most pupils could observe and draw the cells (labelling individual cells), others who are progressing less well may need extra support in identifying what it is they are supposed to be observing. Where in-class support is available, it is important that the person providing the support is able to use a microscope and identify cells themselves, which may require some staff training.

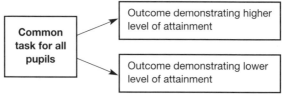

Differentiation by outcome

DIFFERENTIATION BY TASK

On occasion, it may be necessary to set different pupils different tasks, or for them to be directed to different sections of a composite task, such as answering a set of questions about cell structure and function. Some pupils who have progressed further than others may be asked to omit the first three questions which ask basic questions about cell structure. They would then concentrate on the remainder of the questions which become increasingly demanding. Pupils who have not made as much progress may be required to begin at the beginning. It is important, however, in differentiating by task, not to prevent pupils from demonstrating attainment where they can. Excluding pupils from a more demanding task may limit the level of attainment they might otherwise be able to reach. Ensuring that each pupil is suitably challenged means monitoring their learning through a variety of means such as oral questioning, marking and observation.

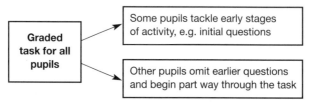

Differentiation by task (graded activity)

It may be appropriate at times to set completely different tasks for groups of pupils. This has to be done sensitively, to avoid negative labelling of some pupils. An example could be if you ask some pupils to produce drawings from photomicrographs, to reinforce both their understanding of cell structure and their ability to interpret images of cells. At the same time, other pupils who seem to have a firm grasp on cell structure might be required to use secondary sources to find out more about the sub-cellular structures such as the nucleus or cell membrane.

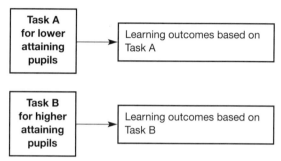

Differentiation by task

When you are providing appropriate in-class support for lower attaining pupils, it is important not to forget the more able and the 'gifted and talented' pupils. They may need extension and enrichment activities which allow them to demonstrate higher levels of attainment. Although this can often be achieved through careful planning for differentiation, by task and by outcome, be aware of where some pupils might benefit from enrichment activities. A fall off in pupil motivation can follow from insensitive 'lock step' approaches which fail to challenge more able pupils appropriately. Such pupils are often able to study a science topic in more depth and breadth than other pupils, and can grasp theoretical ideas more easily. They may also be able to demonstrate their learning in more adventurous ways, such as through poster displays, *PowerPoint* presentations, audiovisual approaches, or even by constructing web pages. However, as motivational tools, these presentation approaches could also be used with other pupils to good effect.

Progression in ideas

The *Framework for Teaching Science* provides a good road map for planning for pupil progression in terms of the key concepts, of which 'cells' is one. The end of year objectives provides a framework for ensuring that pupils' knowledge and understanding of the cell concept is developed in a variety of ways, including depth, breadth and context.

PROGRESSION IN PUPIL UNDERSTANDING OF THE CELL CONCEPT

Obviously it is vitally important that pupils' understanding of the cell concept, a key idea, is thorough. Many pupils may fail to distinguish between a range of words and concepts used to describe 'small things', such as cells and

THE CELL CONCEPT	DEPTH OF UNDERSTANDING OF CELL STRUCTURE AND FUNCTION	BREADTH OF UNDERSTANDING – TOPICS REQUIRING UNDERSTANDING OF THE CELL CONCEPT	CONTEXTS WHERE CELL CONCEPT IS IMPORTANT
Year 7	Cell structure Cell growth Specialisation Cells, tissues and organs	Fertilisation	Sex education
Year 8	Cells obtain energy through respiration Enzymes	Human nutrition Digestion Respiration and gas exchange Micro-organisms Fighting infection	Health education Biotechnology Disease Immunity
Year 9	Equation for respiration Genes	Healthy bodies Reproduction Selective breeding Photosynthesis Plant structure and function	Substance use and abuse Agriculture

molecules. It is possible that they may interpret the statements that 'all living things are made of cells', and 'proteins are required for growth' to mean that proteins are made of cells – a confusion over 'cell' and 'molecule' perhaps. Following work done on cells, using microscopes and other images of a range of cell types, it may be useful to look at the chemicals which make up cells. This may be an appropriate stance to take when looking at nutrition in Year 8. It may also be useful to teach cell chemistry when pupils have covered the particle model for matter in Year 8, where they will meet atoms, molecules, elements and compounds. This is a good example of where two of the key ideas (in this case 'cells' and 'particles') overlap: obviously there is no single, linear approach to mapping out the progression of pupil understanding. Many ideas depend on the grasp of other, related concepts. This is what is sometimes meant by the spiral curriculum, where concepts are revisited in the light of other learning, and then refined, broadened or recontextualised. However, the *Framework* does provide a good indication of what needs to be covered and when, so that revisiting is effective, efficient and not just repetition.

Contemporary science

Much work on cell biology exists at the frontiers of science. Whilst our understanding of basic cell structure and function are not recent developments, a lot of work is going on in the worlds of industrial and academic research looking at the applications of cell functions. These include biotechnology, where we seek to obtain useful substances from the products of living cells, and in cloning.

At this level, pupils do not need to know details of biochemical processes. However, they should be taught about the uses to which some cells are put in terms of the production of useful substances. In unit 8c pupils learn about basic biotechnology, including the production of antibiotics. This provides an opportunity to discuss the pharmaceutical industry – not in terms of detailed processes, but how scientific research is being carried out to discover new medicines and drugs. Contemporary issues could also be dealt with here, such as the problems caused by the overproduction of antibiotics over the past 20 years, which is a key driver in the search for new antibiotic substances.

Teaching about microscopy could include reference to new approaches in optical and electron microscopy, including the use of computers for image processing. Many web sites contain descriptions of such developments.

Ideas for activities

This example shows an approach which might be used to support a bought-in scheme. The section on electromagnetism overlaps with the forces unit but fits in here well. Through completing this unit on electricity, pupils should build upon their prior learning of electricity at Key Stage 2.

Popular misconceptions when covering electricity are mostly based around the concepts of voltage and current. Voltage is the push that makes electricity flow whilst current is a measure of the actual flow. The understanding of the concept of voltage and current is key to making good progress in Key Stage 3 physical processes.

LEARNING OBJECTIVES	TEACHING ACTIVITIES	LEARNING OUTCOMES	POINTS TO NOTE
Know about the background to our current knowledge of electricity	Pre-test activity as described in Chapter 1, or spider diagram based upon what we know and would like to know about electricity **Assessment for learning**: should establish a starting point for the unit	Know about the history and uses of electricity	Homework activity: list as many electrical devices as possible, under headings of labour saving or entertainment devices.
Understand that electricity provides energy to make things work Understand that a complete circuit is needed for energy to be transferred	Fault finding in a range of different circuits	Show understanding of circuit ideas by successfully finding faults Make and explain predictions about circuits	
Know that current is measured in amps Understand that putting more lamps into a circuit decreases the current Understanding that increasing voltage increases the current	Pupils construct circuits with one bulb and measure the current in different sections Find the effects of changing number of bulbs and the voltage	Describe how increasing the number of bulbs reduces their brightness in a series circuit Know that current does not change in a simple series circuit, that it is not 'used up' as it travels in the circuit	This activity enables pupils to observe the qualitative effect of changing voltage on the brightness of the bulb. **Misconceptions**: care is needed here to dispel popular misconceptions relating to understanding of current and voltage.
Understand voltage and its measurement	Following an initial demonstration, pupils investigate altering the voltage and its effect	Identify a cell or battery as a source of energy	Pupils should now make measurements in a quantitative way by using voltmeters and ammeters.

LEARNING OBJECTIVES	TEACHING ACTIVITIES	LEARNING OUTCOMES	POINTS TO NOTE
Understand that there is a voltage wherever energy is entering or leaving a circuit.	upon current in a quantitative way Pupils interpret and show their results	**Numeracy**: data handling and construction of graphs	Some comparison of how pupils showed their own results would be a nice forum for discussion.
Distinguish between energy and current Model ideas about circuits	Using observations from previous work, draw together the main ideas Introduce the idea of a model as 'imagine what it is like...' Discuss a range of models to explain ideas about circuits, which may include shopping trolleys, pumped liquids	Distinguish between current and electricity Explain current in terms of a model See that models help us to understand the world around us **Modelling**: pupils can be given the opportunity to develop their own concept models for electricity flow	It is essential to spend time on this activity. Pupils will come up with some interesting analogies for their own models. This can seem heavy going at times but also there are moments when you can see the light bulbs switch on in the pupils' heads: 'Yes! I now understand.' Well worth persevering with! A useful video is available from BBC's *Science in Action* series.
Know about series and parallel circuits	Pupils measure current and voltage, whilst comparing the brightness of lamps in a range of parallel circuits Make comparisons of series and parallel circuits, using circuit diagrams	Predict and explain measurements of current in different parts of parallel circuits Use a model to explain observations Identify strengths and weaknesses of a model for electricity	Use DART activity to support learning.
Know about electromagnetic effects	Use low voltage lab packs, insulated wire, nails and paper clips to investigate electromagnetism Investigate the effect of changing the number of coils, current	Make and explain predictions about circuits Provide descriptions and explanations for the effects of electrical current	It is useful to have some plotting compasses to hand so that pupils can investigate the magnetic field for themselves Use the *Science Bank* video on electromagnetism to support pupil experiences.
Review	Use assessment quiz to find out what pupils do and do not know	**Assessment for learning**: opportunity as described in Chapter 1	Opportunity here to fit in 'personal capabilities' (see Chapter 1).

Key issues

Misconceptions

Misconceptions can happen at both the levels of the teacher and the pupil. For teachers it is essential that you are familiar with the content that you are expected to teach. Electricity, in particular the concept of current flow, is one of the areas of physics that strikes fear into some non-specialists. The best way to tackle this issue is through having a 'hands-on' departmental meeting where the physics specialists within the team demonstrate the key concepts behind their teaching of Key Stage 3 electricity. Another possible strategy is for staff to identify what they believe to be the pupils' stumbling blocks and misconceptions. From here the department can work together on trying to clarify any misunderstandings. This model can be repeated for different aspects of science to help to clear up other misunderstandings in the curriculum.

The problem areas of the electricity topic are based around the understanding of current and voltage. In dealing with the issues of current and voltage much of the difficulty arises from pupils struggling to separate the two distinct concepts and how they relate to each other. This is where the role of the teacher as a good storyteller comes into its own: there are many established and powerful analogies that can be used to explain the phenomena. These include hosepipes and water flow, steeplechases, trains and trucks on railway loops, even shopping trolleys. If you take the time to develop your own analogy then why not transfer your trust in this skill to your pupils. Given a suitable amount of time your pupils will enjoy trying to develop their own analogies and models. The starter activity to a lesson such as this would be for you as teacher to talk through your favourite story about voltage and current flow. Following your description pupils then have to develop their own ideas either individually or working in pairs. Then, regrouped into fours, they are given chance to share their ideas and then choose their own favourite model. The favourite models are then further developed by the teams of four and presented back to the whole class as plenary session.

Numeracy

There are a variety of opportunities presented in the electricity scheme of work where numeracy can be tackled. These include using data gathered from investigations such as the change in current flow as voltage is varied. The data can then be used to construct line graphs. Pupils should then be set the task of interpreting their own results.

A more complex numeracy issue is presented in attempting to interpret data relating to series and parallel circuits. Classes can either gain their own data through investigating series and parallel circuits using voltmeters and ammeters. Or they can be provided with pre-generated data to interpret for themselves. The challenge of interpreting data from series and parallel systems is that although data from series circuits are easy to interpret (current staying constant throughout the system), in parallel systems there is a division of current throughout the branches of the circuit. This data can be quite challenging for pupils to interpret, and will help them in their analysis and evaluation of circuit diagrams.

The solar system and beyond

(Year 7, from QCA unit 7L)

Ideas for activities

Through completing this unit on the solar system and beyond students should build upon their prior experiences of space at Key Stage 3. Popular misconceptions when covering space are mostly based around the fairly simple concepts of day and night and the relative movements of sun, Earth and moon.

This topic provides an ideal opportunity for pupils to complete some project work once they have established some common foundation activities. In this unit several published schemes are identified as providing appropriate resources for specific activities.

LEARNING OBJECTIVES	TEACHING ACTIVITIES	LEARNING OUTCOMES	POINTS TO NOTE
Learn about the history behind peopled space exploration	Pre-testing as described in Chapter 1 Followed by activity relating to the history of space flight	Find out what they already know	History of space flight gives opportunity to make a time line.
Learn about the solar system	Space scavengers activities Students to work in groups to put together some form of presentation, visual or otherwise, based on gleaning information about the planets that make up our solar system	Find out about the solar system **Continuity:** An ideal transition activity between KS2 and KS3	This should be completed using the varied media resources of a learning centre or library.
Learn about light sources – the sun and stars	Activity 3.3: 'Lighting our way' Folens 'Galaxies' and 'Twinkle Twinkle' activities also good to support this	To find out about sources of light and how our planet is relatively small in the big scheme of things	
Model sun, Earth and moon relationships	Pupils use three different sized balls and a source of light to describe the relative movements of the sun Earth and moon, night and day, phases of the moon and the two main types of eclipse Alternatively use activity 3.5 from *Eureka*	Pupils should understand a variety of situations having modelled them and attempted to describe them for themselves, then make descriptive diagrams to show phenomena	This can be an area for strongly held misconceptions. A good point to use modelling. Use a lab with blackout.

LEARNING OBJECTIVES	TEACHING ACTIVITIES	LEARNING OUTCOMES	POINTS TO NOTE
Understand satellites and gravity (two lessons)	Using materials and activities 3.7 directly from *Eureka*	Find out more about our solar system **Learning resources:** opportunity to use wide range of media, e.g. CD-ROMs, video, Internet	**Numeracy opportunity:** using data, calculating and graph type work.
Revision (one lesson)	Show solar system video and answer associated questions related to the video Complete some self test and revision activities as described in Chapter 1	Revision activity **Assessment for learning:** opportunity as described in Chapter 1	

Key issues

Learning resources

This unit provides an ideal opportunity to use a variety of media resources. The nature of the solar system and beyond topic lends itself to establishing a mini project. By combining two sections of a topic into one you can create time in which to challenge your pupils to work closely together in small teams on their personal capability skills (see Chapter 1) by completing their own mini project. By using a learning centre or school library the project can offer an alternative to laboratory based activities and create a great atmosphere in which to have a space programme challenge. Emphasise that as a teacher you will be looking at their presentation skills. One way of doing this is to challenge your pupils to create a visual display showing one part of the solar system, and to give a mini presentation on what they have found out. Much of the first lesson in a project such as this will be spent establishing defined groups and clear goals and expectations for each of these groups. It is worth taking time here to think about what groupings you wish to establish. For some classes friendship groups will work but in many instances it will be necessary to establish groupings that share a range of skills. In a learning centre environment students can be given time to research, using all the available media resources of books, CD-ROMs, Internet and maybe video material too. Once they have done their research, students should be given further time and materials to prepare their presentations. We suggest one sheet of A1 paper and access to lots of creative arts materials, plus computers and printers if possible. Finally pupils feed back their findings to the whole class by being given a fixed time period in which to describe their poster work. If the activity is repeated throughout a whole year group, it is possible to establish a rolling ever-changing display in some prominent part of school in which to show off not only the skills of your pupils but also the teaching flair of the science department too. One thing to consider is that following the presentation lesson students will need to be given either some homework relating to the solar system as a whole or a set of notes preferably drawn from their own findings.

Progression

The very things that make the solar system and beyond topic attractive to pupils at Key Stage 3 and such an opportunity for a mini project also apply to Key Stage 2 pupils. With a little more time and effort invested it is possible to create your own bridging project whereby you gain better links with your primary feeder schools. It could be that at the end of their Key Stage 2 courses the pupils are challenged to gather information that will then be used in their new secondary school to create a display, using the secondary school learning centre. Pupils could also be provided with exercise books from the secondary school in which to record their findings. Some primary schools would be happy to provide details of Key Stage 2 outcomes on the inside cover of the books. Being given an exercise book is a simple gesture yet, as long as some work has been completed within its covers it provides a genuine transition working document from primary to secondary school. Alternatively primary school pupils can be challenged to create display materials as in the above 'space programme,' that will go up on the walls of secondary schools giving them some feeling of ownership and becoming part of their new school in the new term. On arrival at secondary school individual students could be challenged to complete a profile in their transition books of the whole solar system by gathering data from the posters on display.

Atoms and elements
(Year 8, from QCA unit 8E)

Ideas for activities

In this unit pupils learn to use a particle model to describe what happens when elements are combined to form other materials. They learn that elements are composed of one type of atom and that elements can be grouped by their properties.

This unit follows on from unit 7G 'Particle model of solids, liquids and gases', although some teachers may choose to introduce pupils to particle ideas through their work on elements and compounds. The unit provides a range of pupil activities including researching elements, recording observations, looking for patterns, modelling reactions and planning how to investigate substances safely.

LEARNING OBJECTIVES	TEACHING ACTIVITIES	LEARNING OUTCOMES	POINTS TO NOTE
Know that there is a huge variety of materials Understand that there are a small number of elements from which all other materials are made	Classification exercise with a wide range of materials to separate elements from other materials	Name a wide variety of materials Describe elements as the materials from which everything else is made	Include some gases in the sorting exercise. **Misconceptions**: some pupils may find difficulty in understanding the hierarchical nature of classifications ('it cannot be an element because it is a metal').
Distinguish between elements and other materials Understand that each element is made up of one sort of particle and these are called atoms	Modelling to show how elements may be combined to form other materials Use particle model Introduce chemical symbols to represent elements **Assessment for learning**: check pupils can distinguish between elements and other materials, and between atoms and molecules **Continuity**: particle model was introduced in Year 7, but may not have touched on ideas about different sorts of particles	Recognise the symbols for some elements Demonstrate an understanding of the relationship between elements and atoms and between elements and non-elements by the use of drawings	**Modelling**: some teachers may wish to make explicit the limitations of models used. Models are a useful tool for explaining difficult scientific ideas, but pupils need to be made aware of their status and limitations. **ICT**: provides an effective way of visualising particle models.

LEARNING OBJECTIVES	TEACHING ACTIVITIES	LEARNING OUTCOMES	POINTS TO NOTE
Know that elements vary in their appearance and state	Research, using ICT sources or printed materials, into the properties of different elements. Display the results to show how elements can be grouped by their properties, linked to their position in the periodic table **Learning resources**: check web sites and CD-ROMs for accessibility by Y8 pupils as well as relevant content. Posters are available from chemical and mining companies	Describe some differences between elements Make some generalisations about elements	**ICT**: web sites and CD-ROMs are a possible source of information for this activity. Pupils would need to be shown how to search, select and record the information they need. **Literacy**: some pupils may need to be taught how to extract useful information from a range of resources.
Understand that molecules of compounds are formed when atoms of elements combine Represent and explain chemical reactions by word equations, models or diagrams	Record observations of some reactions between different elements to form compounds Represent the reactions by equations, models and diagrams	Describe what happens in some chemical reactions and name the product Explain compound formation in terms of atoms joining	Links to unit 7F. **Literacy**: representing elements as symbols and reactions by equations are important aspects of 'scientific literacy'. Establishing clear ground rules for their use may help avoid misconceptions.
Predict what might be formed from a chemical reaction between two elements	Predict, observe and explain the effect of heating certain metals in air	Predict the product of some simple reactions Interpret the names and/or formulae of binary compounds in terms of the elements of which they are composed	**Misconceptions**: pupils may believe that heating in air or burning results in a reduction in mass as material is 'used up'. Also, gases may be considered to have no mass or negative mass. This leads to problems in understanding that the formation of metal oxides results in mass increase.
Distinguish between elements and non-elements Explain how their work provides evidence about a question	Plan and carry out a practical activity to collect evidence to show whether an unknown substance is an element **Ideas and evidence**: explain the importance of scientific understanding in formulating a question to investigate	Identify methods that will provide appropriate evidence Explain how their results provided evidence	Some pupils will need considerable help in planning this activity. **Scientific investigation**: this activity helps pupils to identify key ideas about elements and compounds, and to use these to plan their own investigation. They should consider the weight of evidence before arriving at a scientific conclusion.

Key issues

Assessment for learning

In this context assessment *for* learning is a description of the teaching activities that provide information about levels of understanding that in turn promote further learning. These activities differ from summative assessment *of* learning tasks:

- they should be quick and easy to administer;
- they occur during teaching, not after teaching;
- they may provide qualitative or quantitative information;
- they can be fun, providing they are not seen as part of a testing regime.

Particle theory is one of the five key scientific ideas identified in the science strand of the Key Stage 3 Strategy. It forms a 'conceptual building block' for the understanding of many other scientific concepts, so it is worth spending some time to check on pupils' understanding as they progress through units such as unit 8E on atoms and elements. This will help pupils to recognise the progress they are making, and also give you a chance to modify your short-term teaching plans in response to their needs.

Below are three suggestions for how pupil understanding may be assessed. These ideas are not original, but they do work as classroom activities. As with any suggestions, there are pros and cons with each; adapt them for your own situation or use them as a stimulus for developing your own materials.

1. OPEN-ENDED QUESTION

Pros: easy to administer and provides information about individuals.
Cons: needs marking; may be treated as a test by pupils.

2. 'TRAFFIC LIGHT' DISCUSSIONS

The pupils are given three pieces of card – one red, one green and one yellow. A statement is read out and after two seconds thinking time all pupils in the class have to hold up one of their cards.

Red means 'I disagree'. Yellow means 'I am not sure'. Green means 'I agree'.

Pros: quick, easy way of getting a snap shot of the class.
Cons: information provided is superficial and may be inaccurate.

Example statements:

- All elements contain one sort of atom.
- All compounds contain two sorts of atom.
- Compounds are pure substances.
- Particles in a gas are smaller than particles in a liquid or a solid.

Example of open-ended question

The diagrams represent different gases.

The atoms of different elements are represented by ○ and ●

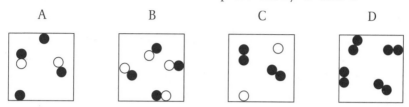

Which of the four diagrams, A, B, C or D, represents a single compound? ☐

Give a reason for your answer:

Which of the four diagrams, A, B, C or D represents a mixture of two elements? ☐

Give a reason for your answer:

For other statements concerning particle theory, or other uses of 'traffic lights', see Science Year CD-ROM materials.

3. Practical activities

Short practical activities, perhaps arranged as a 'circus' around the laboratory, can be used to stimulate pupils' curiosity and encourage them to discuss their ideas and explanations. I would arrange three or four activities, each to be tackled by groups of up to four pupils for no longer than ten minutes each. Pupils carry out the practical task and then write their explanation for what they see, before moving on to the next activity.

Pros: provides first hand experiences which can help to take pupils' understanding forward.
Cons: written responses need marking, and so immediacy of feedback can be lost unless practical tasks are then discussed by the class.

Example activities:

• Comparing the compressibility of air with water in large sealed plastic syringes.
• Floating a (blunt) needle on water using a piece of filter paper.

- Comparing the magnetic properties of iron sulphide with iron filings and sulphur.

FINALLY...

- This form of assessment does not lend itself readily to providing a score or mark.
- This is a good technique for identifying misconceptions held by the class.
- Many of these activities provide a useful starter activity for a lesson as pupils are usually keen to discuss their responses.
- To promote learning the results of these assessments should influence short-term lesson planning.

Plants and photosynthesis
(Year 9, from QCA unit 9C)

Ideas for activities

In this unit pupils learn about photosynthesis as the key process for producing biomass in plants. They learn about the ways plants obtain raw materials and energy to drive the reaction, and about the importance of the reaction to humans and the environment. The unit draws on learning from many other areas of the science curriculum, and leads on to unit 9D, Plants and food.

This unit provides many opportunities for practical work and for handling living material. However, teachers may find computer simulations useful to support their teaching as some of the practical work can produce variable results.

LEARNING OBJECTIVES	TEACHING ACTIVITIES	LEARNING OUTCOMES	POINTS TO NOTE
Learn about the process of photosynthesis Understand how plants obtain the raw materials required Know how to represent the process as a word equation	Use a range of fresh plant material such as seedling, herbaceous plants and woody twigs to teach about plant growth Photosynthesis as a chemical reaction Using secondary sources of data look for patterns in changes to the atmosphere surrounding a leaf **Numeracy**: interpreting graphs and data to identify trends and patterns. Judging the reliability of data	Identify sources of raw materials Explain the terms 'photosynthesis' and 'biomass' Construct a word equation Explain how carbon dioxide levels are related to light levels	**Misconceptions**: this complex topic provides many areas where misconceptions can interfere with learning. The most obvious is the notion that plants feed from the soil, though the complex language, the nature of evidence and the nature of the reactions also cause pupils difficulty.
Understand the role of chlorophyll and light Test a leaf for starch Investigate rate of photosynthesis	Test a variety of leaves for starch Design and carry out an investigation into the effect of changing light intensity on photosynthetic rate **Science investigation**: this is a challenging practical activity. Use of a simulation experiment may provide a more reliable approach to developing investigative skills	Relate the presence of starch in a leaf to chlorophyll distribution and light exposure	Some of this activity links to work in Y8 on respiration. **Continuity**: this is the first opportunity in the QCA scheme of work for pupils to learn about photosynthesis, although it clearly links to other units on feeding and respiration. **ICT**: oxygen concentration and light intensity may be monitored with sensors.

LEARNING OBJECTIVES	TEACHING ACTIVITIES	LEARNING OUTCOMES	POINTS TO NOTE
Know that glucose is used in the production of other substances Learn about increase in biomass Understand that photosynthesis provides energy for respiration	Remind pupils how glucose may be used for respiration and how the structure of sugars and starches is related, by burning in oxygen and testing the products of combustion Investigate a range of economically important plant products using secondary sources	Understand the role of photosynthesis in respiration, the production of biomass and synthesis of other materials	This provides an opportunity to consider social and environmental issues linked to plant growth and exploitation. **Learning resources**: for this activity include video, paper based and digital media as well as the use of artefacts.
Understand the functions of roots Learn about uses of water Understand plant mineral nutrition	Use living material, slides and video to learn about root structure, function and adaptations Interpret experimental evidence of water movement through the plant Use secondary sources to learn about the need for minerals to support plant growth **Literacy**: DART activities would be appropriate to help pupils understand the material provided	Explain the functions of water in a plant Explain the structure, function and adaptations of roots Understand why mineral salts are needed for healthy plant growth including the use of nitrogen in production of protein **Ideas and evidence**: Van Helmont's experiment	Good opportunities for the use of a variety of living material for pupil observation and investigation. Opportunities for links with work on diet and healthy eating. **ICT**: a wealth of information exists on the Internet on healthy eating and the role of minerals. ICT provides a variety of methods for pupils to present their work.
Understand the importance of green plants in the environment, including maintaining the balance of atmospheric gases, and the links between respiration and photosynthesis	Investigate global changes in oxygen and carbon dioxide levels over time through the use of secondary sources **Modelling**: the uses and shortcomings of using projected figures to predict global change Debate some of the issues around forestry conservation and land clearance **Literacy**: helps pupils to develop listening as well as speaking skills	Understand how photosynthesis and respiration influence the balance of gases in the atmosphere **Assessment for learning**: explicit links to be made between this review activity and the introductory activity. Learning outcomes from the first lesson used in planning the details of this lesson Learn about the advantages and disadvantages of forest conservation from different perspectives	Links with greenhouse effect and global warming in unit 9G. **Contemporary science**: pressure groups publish considerable amounts of conflicting evidence that may be useful.

Key issues

Misconceptions

Pupils (and some teachers) find photosynthesis a difficult area of science. The problems mainly arise from three different areas.

- It covers a range of levels of organisation, from molecular reactions, through cell and leaf structure, right up to influences on global climate change.
- Many science concepts are met, including energy transformation, diffusion, adaptation and nutrition.
- The technical language used in this topic can present its own problems in spelling, pronunciation and understanding.

The topic is met for the first time in Year 9 in the QCA Scheme of Work. At this stage in a pupil's scientific education, teaching draws on many other areas of the science curriculum:

- Unit 7I Energy resources;
- Unit 8A Food and digestion;
- Unit 8B Respiration;
- Unit 8D Ecological relationships;
- Unit 8F Compounds and mixtures.

Photosynthesis also draws on four of the five key scientific ideas identified in the science strand of the Key Stage 3 Strategy (cells, interdependence, energy, particles and forces). It takes skillful teaching to identify in which of these areas a pupils' understanding is less secure and then to take appropriate action, but for me this is why a teacher's knowledge of misconceptions is so important.

By the end of the topic some children will have made more progress than others, so it is worth finding out which misconceptions remain. This will give you useful information for planning future lessons, revising your teaching materials, and planning the Year 10 curriculum.

MISCONCEPTIONS ABOUT PHOTOSYNTHESIS

The following activity should help you begin to think about the common misconceptions held by pupils regarding photosynthesis.

A Year 9 teaching group has just completed a series of lessons on photosynthesis. The teacher has decided to set a task to encourage the pupils to review their understanding of the topic, and to allow the teacher to assess the learning that has taken place. She asked them to work in small groups and produce a poster that summarised their understanding of photosynthesis. The title they were given was 'How do plants feed?'.

The statements below are transcripts of the text on their posters. (In each case the text had been supplemented by one or more graphical image.) Read through the text from each poster, deciding which shows the most complete understanding of the topic. You may like to rank them according to the level of understanding shown.

POSTER	TRANSCRIPT OF TEXT	RANKING
1	Plants breathe in carbon dioxide. Plants make carbohydrates which contain carbon, hydrogen and oxygen. The roots absorb water from the soil. Plants take in energy from the sunlight and then energy comes out when they feed.	
2	Carbon dioxide, water and light all mix together to make food and oxygen.	
3	The light comes down and the green on the leaf soaks it up. The green colour is chlorophyll. In this is starch. Starch is needed to absorb light.	
4	Plants need light and chlorophyll to make starch. They use there leafs to absorb light from the sun. They then turn the energy they got from the sun into starch. This process is called photosynphisis.	
5	It needs sunlight, water, carbon dioxide to stay alive. And then it makes its own food.	
6	Plants need light and energy and lots of water. They need all these things so they can grow and produce more plants.	
7	Plants convert sunlight into food using photosynthesis. The green pigments called chlorophyll is where this occurs. Carbon dioxide is taken in and used to make the plant grow and oxygen is given out by the plant.	
8	Water and sunlight are both needed to grow plants. The water gives the plant it's oxygen and it's hydrogen.	
9	Plants need sunlight, carbon dioxide, H2O and chlorophyll to make starch. Plant use the green part of the plant to absorb the energy from the sun to build up the starch.	
10	The leaf of the plant absorbs the light and produces food. It is also taking in carbon dioxide. The roots take in the goodness from the soil. Also known as vitamins and minerals. The plant also takes in the moisture from the rain, this keeps them alive. In all plants photosynthesis takes place. This is when the leaf takes in the suns light and produces energy.	

From this sample what would you say are the most common misconceptions held by this class? I have listed some misconceptions about plant nutrition in the table below, but you may identify others (there is space to add these to the list).

MISCONCEPTION	NOTES	EVIDENT IN SAMPLE STATEMENTS?
Technical language	Words such as photosynthesis and chlorophyll are used inappropriately	
Everyday language	Key words are used in an everyday context rather than a specifically scientific context. Examples include feeding, nutrients, breathing.	
'Energy'	Photosynthesis can be viewed as an energy transformation from 'light' to 'chemical'. Some pupils do not regard light as an energy source, or food as a store of chemical energy.	
'Jigsaw' understanding	Pupils may have several clearly understood pieces of information, but are not sure how to fit them together into a coherent picture of the overall process.	
'Growth and living'	Is growth associated with an increase in (organic) mass or in height? Does feeding always result in growth and survival? Is staying alive or growing an indication that feeding is occurring?	
Biological molecules	Confusion sometimes exists between 'glucose', 'starch' and 'carbohydrate'. (This and some other misconceptions may be reinforced by some of the practical work associated with teaching this topic.)	

FINALLY...

- Pupils already have ideas and explanations about topics before they meet formal scientific teaching.
- Some of these ideas can interfere with them developing a scientific understanding. These ideas I describe as 'misconceptions'.
- Pupils do not always develop the expected understanding from a teaching activity.

- It is worthwhile spending some time finding out what pupils know about a topic both before and after formal teaching.
- Misconceptions are not easily changed. A more achievable target for a teacher is to help a pupil become aware of their own misconceptions, and see how these compare with scientific explanations.

Environmental chemistry
(Year 9, from QCA unit 9G)

Ideas for activities

By completing this set of activities, pupils will be building on their prior knowledge of two areas of science: (i) acids and alkalis, (ii) rocks and soils. The work should allow them to understand the connection between the two and the topic offers an ideal opportunity for pupils to apply their knowledge to current issues such as pollution.

Areas of misconception tackled in this topic include confusion about the greenhouse effect and holes in the ozone layer. Pupils are offered the opportunity to think 'like scientists' by researching thoroughly and considering all of the evidence before attempting to reach any conclusions.

There are opportunities for small group work and investigative practical work as well as plenty of occasions for evaluating secondary sources.

LEARNING OBJECTIVES	TEACHING ACTIVITIES	LEARNING OUTCOMES	POINTS TO NOTE
Investigate whether soils are different from each other and how the pH of soils affects which plants will grow Use prior knowledge of acids and alkalis to suggest ways of changing soil pH	Recap knowledge of acids and alkalis from Y7 Discuss soil formation and different types of soil (from KS2) Use seed packets, etc. to show that different plants require different types of soil Test different types of soil using just pH paper/solution and then using a soil testing kit – tabulate results	Produce a table (in rank order) showing the pH of different soils Make suggestions for changing the pH of soils that are strongly acidic or alkaline	**Safety issues:** consider the use of various chemicals and the handling of soil **Continuity:** pupils will have studied plant growth and nutrition at KS2, which could provide a link to looking at environmental impacts on growth. They will also have studied acids and alkalis earlier in KS3.
Understand how rocks and building materials change over time and what has caused these changes	'Spot the difference' activity using weathered and non-weathered or before and after pictures of landscapes and buildings; pupils list the differences and suggest reasons for these differences	Understand that materials do change over time and that a combination of factors can be responsible for these changes	
Know about the causes of acid rain Understand the effects of acid rain Learn about reducing acid rain	Investigations: two experiments (half the class to carry out one and half the other) 1. Effect of dilute acid on a range of rock samples	**Science investigations:** these practical activities allow pupils to carry out fair tests (with relatively low level procedural understanding) whilst tackling more	This work offers opportunities for a number of key skills such as: – Communication (writing a report);

111

LEARNING OBJECTIVES	TEACHING ACTIVITIES	LEARNING OUTCOMES	POINTS TO NOTE
	2 Effect of dilute acid 'rain' on the germination and growth of seeds, e.g. cress	demanding concepts in the underlying science knowledge and understanding. This is particularly so in the pH and germination investigation.	– Application of number (presenting and interpreting results from investigation);
	Results presented to the class		– Thinking skills such as planning an investigation.
	Examine video or still photograph evidence of the effects of acid rain on plants and animals	**Differentiation:** Investigation 1 provides lower level demands on conceptual understanding than investigation 2.	
	Research the sources of acid rain and using this and the video/pictures write a newspaper report about the issues including ways of reducing the problem	Present results to peers	
		Develop research skills and creative writing	
Learn about air and water pollution Consider how pollution can be monitored and controlled	Brainstorm causes of pollution and ask pupils to rank the causes in order of importance	Evaluate evidence in order come to a decision	This exercise shows that there may not be a definite or agreed answer to a scientific question, it is more a case of understanding the strength of the evidence.
	Using a local or named location, the pupils research the changes in levels of pollution over the last 200 years		**Ideas and evidence:** this provides opportunities to consider how scientists work to monitor the environment and how the evidence for climate and environmental change needs careful consideration.
	Small group presentations to the class		
	Pupils submit proposals for how pollution could be controlled		**ICT:** use of Internet for research work and the presentation of findings.
	Teaching and learning: use of small group work and pupil presentations.		Opportunities for:
	Environmental monitoring could be carried out locally using pH sensors to monitor air or rain water and oxygen sensors to monitor water quality		key skills – working with others and problem solving;
			thinking skills – reasoning;
			citizenship – education for sustainable development and other environmental issues.

LEARNING OBJECTIVES	TEACHING ACTIVITIES	LEARNING OUTCOMES	POINTS TO NOTE
Consider the question: is global warming happening?	Small groups of pupils given questions/issues to research, e.g. is the Earth warming up and if so, what is the evidence for this? Each group presents to the class Whole class discussion Each student summarises the issues from a personal point of view (possibly for homework)	Interpret data and other sources of information Draw conclusions based on the evidence	**Misconceptions**: need to clear up any confusion between the greenhouse effect and holes in the ozone layer. **Teaching and learning**: the unit also provides opportunities for: – creative thinking skills; – evaluation skills; – working with others.

Links with other curriculum areas

Acid rain, pollution and global warming:

- Citizenship (topical political and social issues);
- Geography (ecosystems, population and sustainable development).

Key issues

Literacy

Literacy is the key to improving pupils':

- communication, thinking and learning;
- performance in reading, writing and speaking and listening;
- independent learning;
- self esteem and their performance in examinations.

Good literacy is a key factor in raising standards across all school subjects – poor literacy usually means poor performance. Pupils need to read intelligently and express themselves effectively if they are to do well in all their subjects. This does not mean taking extra time during science lessons to teach English skills. What it does mean is that literacy skills can be fostered as part of any subject including science.

It is important that pupils should be able to write using correct spelling, punctuation and grammar, and there is considerable opportunity in the Environmental chemistry unit to practise these skills in the form of report writing, including a practical investigation report. This unit deals with many issues which are currently in the news, such as global warming and acid rain. Newspaper articles can be used to stimulate discussion – helping develop pupils' reading and oral communication skills. Literacy is more than just being accurate on paper. The main medium in school for teaching and learning is the spoken word, and this influences the way that thinking is developed in the subject. For example, pupils need good language skills to explain how an experiment was carried out and the significance of the results. Throughout the unit there are opportunities for pupils to develop and demonstrate their language skills.

It is equally important that pupils can speak precisely and that they can listen to others so that they can respond constructively. They should, in this context, be able to use the technical and specialist vocabulary of science. Equally, they need to be able to recognise, describe, use and apply key scientific ideas to explain abstract phenomena even when they appear in unfamiliar contexts. Speaking can easily be developed by small group discussion, and by requiring pupils to give short presentations based on their research of a variety of topics. By reporting to their peers on their findings from the acid rain work, pupils get the opportunity to demonstrate their oral skills.

It is important, therefore, that science departments pay increasing attention to the development of pupils' literacy skills as a way of improving achievement. This can be simply by displaying key words in laboratories and using writing frames for recording work such as investigations. Teachers must continue to be aware of the specialist language demands of science textbooks and assessments. The unit offers an opportunity to develop (or add to) a scientific glossary the pupils may be compiling with its use of both biological and chemical terminology.

Perhaps the most effective development of literacy skills to support teaching and learning in science can be summarised as involving:

- reinforcement of meaning and use of terminology by pupils in context;
- reducing routine writing such as descriptions of practical work and increasing writing about pupils' own understanding and interpretation of information – creative writing does have a place in science;
- extended writing for other purposes such as 'ideas and evidence' work;
- reading about science issues as well as reading just for information.

Teaching and learning

Science demands versatile teachers and imaginative approaches to bring it to life for pupils and give them a thorough understanding of the subject. Examining key ideas can stimulate pupils' curiosity and help them make connections between different areas of science. Scientific enquiry will link these key ideas with direct practical experience.

The Environmental chemistry unit offers the opportunity to employ some different teaching and learning approaches – in particular, the small group work suggested for the work on pollution and global warming. These are topics on which most pupils have an opinion and may want to know more. By allowing them to do some research in small groups and also become involved in whole group discussion, you give them the opportunity not only to express their own thoughts but to consider those of others.

This demonstrates that the most effective practice does not restrict the principles of scientific enquiry just to 'investigation' lessons. They are integrated into most lessons, even those that involve little or no practical work *per se*. For example, by capitalising on chances in any lesson to encourage pupils to reflect, however briefly, on the evidence that supports scientific interpretations: 'How do you think they measured that?' or 'How would you check the results?'. There are a number of opportunities throughout the unit for pupils to examine the results of others (and their own) more critically.

Effective science lessons always have clear objectives that the pupils understand and are the result of good planning to include:

- high expectations;
- appropriate exposition;
- question and answer;
- monitoring of learning;
- inclusion of contemporary issues and scientific applications;
- class/small group discussion.

Opportunities for all these can be found in the Environmental chemistry unit.

Contemporary science

'there is a tendency for teachers to 'play safe', with content coverage seen as an imperative, and a consequent lack of imaginative teaching. In science departments where teachers have the professional self-confidence to discriminate in their selection of content, building on and revisiting previous learning rather than attempting full coverage, time becomes available for enrichment of the subject. This might include, for example, the searching use of questions to develop deeper understanding, more time considering practical applications of scientific principles or debate about contemporary science.'
(*Secondary Subject Reports 2000/01: Science*, OFSTED, 2002)

In a single century, extraordinary successes in science and technology have changed the world and this world has come to rely on a significantly increased population in science-related employment. Science-related issues play such an important part in contemporary life that science teaching must surely include them. The Environmental chemistry unit offers opportunities to look at how contemporary science benefits society by allowing us to develop alternatives to fossil fuel use. It also allows discussion of the powers and limitations of science in such matters. For instance, there is still controversy over the existence, let alone the cause, of global warming. In particular, pressure groups in the USA, citing the work and views of some scientists, continue to lobby the government to 'go slow' on implementing some internationally agreed environmental policies, such as reducing car exhaust emissions. This can lead to further work on how science is dealt with in the media using any current and controversial issue of the moment.

For science teachers there can be a dilemma: how to present scientific knowledge as established fact one day and then the next day discuss the uncertainty of some scientific knowledge? Another potential problem is the tendency for science lessons to have, sometimes quite rigid 'factual' objectives, so that pupils can gain a better understanding of scientific principles. It is not so 'comfortable' encouraging pupils to consider their responses to issues such as cloning, genetic engineering, mobile phone safety, etc. However, it can be found in the curriculum, embedded in the 'Ideas and evidence' strand of Scientific enquiry.

- Should, therefore, controversial/contemporary issues be dealt with by science teachers?
- Wouldn't they be better dealt with in humanities?
- If science does 'grasp the nettle', how can pupil progress be monitored?

There is no reason why science teachers can not take responsibility for this and there are always ways of assessing pupils. If pupils are to be encouraged to develop the skills of, for example, informed discussion, then these skills need to be given time and value.

Within the Environmental Chemistry unit there are some opportunities for contemporary science issues to be addressed – in particular, the issues of pollution and global warming. The unit suggests some approaches that could be used in each case, although they are not the only ones possible. It is worth noting that within this work there is the opportunity to address some issues of citizenship which can often be found closely linked with contemporary science.

USEFUL WEB SITES FOR CONTEMPORARY SCIENCE

New Scientist **www.newscientist.com**
Particularly the 'Hot Topics' list that has issues currently in the news.

Scientific American **www.sciam.com**
Provides three news items per day for the current month about issues making the news.

Nature **www.nature.com**
Features on groundbreaking science.

Wellcome Trust www.wellcome.ac.uk

Issues such as ethics in science and science research.

Chemistry Societies network www.chemsoc.org

'Byte-size' news items about chemistry in the news.

Institute of Physics www.iop.org

'Physics in the News' section.

Institute of Biology www.iob.org

News section with information about current biology issues.

Professional development in action

We began this book by describing a model for the development of teacher expertise, from novice, through competent and proficient, to expert teacher. If you have worked through this book, or parts of it, carried out the activities and reflected on both what we say, and on your own practice, then you will have experienced some form of professional development. However, reading, reflecting and even discussing issues with colleagues are no substitute for the developmental nature of the act of teaching itself. It is by putting plans into practice that real professional learning takes place.

Making it work in the classroom is key to benefiting from any type of professional development. We have probably all been on courses, particularly the one-off, one-day variety, which were highly stimulating and seemed to address a particular professional need we had. However, it was not until what was learnt on the course was applied to classroom practice that the full benefit from the experience could become realised. It is the same with any programme of professional development, whether it is the Key Stage 3 Strategy, working through this book, attending a taught course, engaging in curriculum development or taking part in the burgeoning on-line CPD world.

So, if you have come across some useful ideas in what we have presented, then now is the time to look at how to implement them. In the increasingly team-based approach to departmental management and development, the next step may be to present your ideas for change to your colleagues in school. This may be a new venture for you, particularly if you are new to the profession. However, in an open and supportive team any good, purposeful and well thought out ideas will always be welcome. Or perhaps you are well versed in instigating change. Your task may be to convince colleagues, to take them with you, in thinking through and then planning for the changes you propose.

So, the key to effective professional development is professional action. But that is not enough in itself. Any developmental activity should be part of a cycle: identify objectives, plan, implement and evaluate against the objectives. Did the new approach to initiating Sc1 investigations work? How do you know? What was it supposed to achieve? Did it achieve this? Has it, to quote ourselves from the introduction 'transformed and strengthened teaching and learning'?

Opportunities for engaging in continuing professional development

CPD isn't just about courses or books. Opportunities to 'transform and strengthen' your practice exist in many modes. Below are just a few examples of opportunities for CPD which you might wish to pursue. One way to

ascertain your development needs is to look at the DfES Teachernet site
www.teachernet.gov.uk/standards_framework/, and follow the links to the
Standards Framework.

TYPE	NOTE	USEFUL WEB SITES
Short courses	Good if well focused and matched to your needs. Many organisations offer these, including LEAs, universities and private companies or consultants.	www.ase.org.uk www.shu.ac.uk/cse www.sfe.co.uk www.rsc.org
Accredited courses	Mainly run by universities. Could be at certificate, diploma or masters level. Some universities offers individual units. ASE also offers an accredited course.	See www.hefce.ac.uk/ UniColl/HE/ for a full list of colleges and universities.
Government initiatives	Including Excellence in Cities, Key Stage 3 Strategy, Education Action Zones and Beacon Schools.	For all these go to www.standards.dfes. gov.uk/ and follow the links you want.
Curriculum projects	Getting involved with curriculum projects can be good professional development, based on trying out new ideas in the classroom.	www.shu.ac.uk/pri www.scienceacross.org/ www.britassoc.org.uk/ the-ba
Working with industry	All areas have organisations that help you make links with local industry and HE departments. These include SETPOINTS and EBPs.	www.setnet.org.uk/ www.nebpn.org
Research-based development	University departments may offer opportunities to carry out research through the Best Practice Research Scholarship scheme. Contact your local HE department.	www.teachernet.gov. uk//Professional_ Development/ opportunities/bprs/

As well as taking part in and benefiting from professional development
opportunities, you should collect evidence of all you do. Keep this in a
professional development portfolio. This should include evidence of CPD, as
well as more day-to-day evidence of compliance with the Standards
Framework, particularly if threshold assessment is relevant to you.
Throughout your career, particularly if you are new to the profession, you will
need to provide evidence of professional development: for threshold
assessment, for performance management and appraisal, and for progression
to Advanced Teacher Status or other promotions. You are likely to have begun
this already. If so, look to what development you have achieved through
dealing with the contents of this book. Have you redesigned your assessment
processes based on tackling some of the issues we outline in Chapter 4? Have
you sought out science updating courses based on thinking about the place of

contemporary science in your scheme of work, as we set out in Chapters 2 and 3? Are you now involved in BAYS Clubs, running CREST Awards or hosting a Researcher in Residence – all suggestions from Chapter 1? If the answer is yes, then you will already be benefiting, in a professional – and we also hope, a personal – way from what we have offered here for your attention.

We hope we have provided you with food for thought, and options for action. Enjoy yourself, develop your professional practice, and participate in the effort to raise standards – and motivation – both for yourself and your pupils.

Glossary

Assessment for learning

Assessment is a mechanism that if used in a supportive way with students, as a self evaluation and diagnostic tool, can encourage students to feel ownership and empowerment towards their own progress through Key Stage 3 science.

Contemporary science

It is important that teachers have an understanding of some areas in which scientific knowledge is developing fast, together with the ethical and social issues related to this science, so they can include these in their teaching.

Continuity

The process that allows students to make progress. Continuity gives students the whole picture of science as it opens up in front of them. Good continuity helps students to build their own perspectives of science. A major issue with continuity is the effect of transition from one key stage to the next.

Differentiation

Planned approaches to teaching and learning which take into account the diversity (of ability, aptitude, attainment, gender, culture, etc.) of pupils in a class or group.

ICT

Information and communication technology is a term found mainly in education. The wider world still refers to it as IT. The advent of the Internet in schools and provision of school networks is having an impact on science teaching, but in this rapidly changing area digital cameras, projectors, video, digital microscopes, data-loggers, whiteboards and other electronic gadgets are also having an impact.

Ideas and evidence

This is about 'how science works' – the interplay between questions, ideas, evidence and scientific explanations, as well as a consideration of science as an activity – 'how scientists work'.

Learning resources

This refers to the range of different media such as those available in school libraries and learning centres, as well as in the classroom or on the Internet. Resources can be digital, paper based, artefacts or audiovisual.

Literacy

Science has its own unique, technical language that pupils must master if they are to communicate well in the subject. While this must be addressed, it should be remembered that no amount of 'techno-speak' will mask a pupil's underlying poor literacy skills. All of the communications skills must be part of science teaching, including speaking, listening, reading, writing and use of media.

Misconceptions

Pupils' understandings and explanations do not always match those of the teacher (i.e. those commonly accepted as being 'scientific'). If the gulf is notably wide then the teacher may label the pupils' ideas as misconceptions. In its widest sense the term can be applied to pupils' understanding of words, concepts, observations and explanations.

Modelling

The ability to explain, using a good model or analogy, is the skill of an effective teacher. Models are representations of the world which provide useful explanations, and can be used where trying to describe actual reality is not possible or useful, such as the waterfall model of voltage or potential difference.

Numeracy

The National Numeracy Strategy, although essentially relating to the maths curriculum, also impinges upon other subjects, perhaps none more so than science. Scientific aspects of numeracy include measurement, using graphs, tables and diagrams, using formulae and basic statistics (e.g. calculating means and class intervals).

Progression

Planning appropriate learning experiences to allow pupils to develop their understanding of science. This includes progression between as well as across key stages. It involves developing pupils' understanding in depth as well as breadth, and also their ability to apply knowledge in a variety of contexts, making connections between different areas of science.

Science investigations

These are planned activities which usually begin with a question, hypothesis or prediction. They involve the collection of evidence and the creation of new ideas based on explaining what the evidence means using scientific understanding. Sometimes explanations represent advances in our scientific knowledge and understanding.

Teaching and learning

There is a wide variety of possible teaching and learning approaches available to teachers. Effective selection of which approach to use depends on how appropriate a given method is in terms of the learning objectives being met, and this also relies on teachers' appreciation of the importance of pupils being actively engaged with their learning.

Adey, P S, Shayer, M, and Yates C, 2nd Ed (1995) *Thinking Science: the curriculum materials of the CASE project.* London: Thomas Nelson and Son.

Berliner, D C (1994) The wonder of exemplary performances, in Mangieri, J N and Collins Block, C (eds) *Creating Powerful Thinking in Teachers and Students.* Texas: Holt, Rinehart and Winston.

Bevins, S (2002) *The Expert Teacher.* Sheffield: Centre for Science Education, Sheffield Hallam University.

Bianchi, L (2002) *Developing personal capabilities through the subject curriculum.* Sheffield: Centre for Science Education, Sheffield Hallam University.

Centre for Science Education (1991) *Active Teaching and Learning Approaches in Science.* London: Collins Educational.

Centre for Science Education (2001) *Acclaim Project.* London: The Royal Society.

Cheshire County Council Education Department (1996) *Moon Colony "Bridging the gap" – Key Stage 2/3 liaison.*

DfES (2002 (a)) *Framework for Teaching Science Years 7, 8 and 9.*

DfES (2002 (b)) *Auditing a subject at Key Stage 3.*

Gardner, H (1993) *Frames of Mind: The Theory of Multiple Intelligences.* London: Harper Collins.

Glaser, R (1984) Education and thinking: The role of knowledge. *American Psychologist*, 39(2): 93–104.

Hodson, D (1993) Rethinking Old Ways: Towards A More Critical Approach To Practical Work In School Science. *Studies in Science Education*, 22.

OFSTED (2000) *Progress in Key Stage 3 Science.*

OFSTED (2001) *Reports 2000/01: Science.*

Osborne, J F and Collins, S (2000) *Pupils' views of the School Science Curriculum.* London: King's College.

Roberts, G (2002) *SET for Success: the supply of people with science, technology, engineering and mathematics skills.* London: HM Treasury.

Wellcome Trust (2001)*Valuable lessons: engaging with the social context of science in schools.* London: Wellcome Trust.

U.W.E.L. LEARNING RESOURCES